One Seed to Another

One Seed to Another:

The New Small Farming

Paul Hunter

Small Farmer's Journal - Sisters, Oregon

One Seed to Another
Paul Hunter
Copyright © 2010 Paul Hunter

Publisher

Small Farmer's Journal Inc.
PO Box 1627
192 West Barclay Drive
Sisters, Oregon 97759
(541) 549-2064
www.smallfarmersjournal.com

Authored by Paul Hunter

First printing first edition 2010

Library of Congress catalog card number –
   ISBN 978-1885210-34

*Front cover photo: Ed Joseph tilling
with two Belgian mares. Photo by Dena Joseph
Back cover: Paul Hunter playing slide guitar.
Photo by Kristi Gilman-Miller*

1. Where We've Been / 9

2. Where We Are / 22

3. Where We're Going / 43

4. New First Principles / 57

5. Steps to Take / 85

6. The Trouble With Farming / 97

I am done with great things and big plans, great institutions and big successes. I am for those tiny, invisible, loving human forces that work from individual to individual, creeping through the crannies of the world like so many rootlets, or like the capillary oozing of water, yet which, if given time, will rend the hardest monuments of human pride.

— William James

Conservation is a state of harmony between men and land. By land is meant all of the things on, over, or in the earth. Harmony with land is like harmony with a friend; you cannot cherish his right hand and chop off his left. That is to say, you cannot love game and hate predators; you cannot conserve the waters and waste the ranges; you cannot build the forest and mine the farm.

— Aldo Leopold

If human vices such as greed and envy are systemically cultivated, the inevitable result is nothing less than a collapse of intelligence. A man driven by greed or envy loses the power of seeing things as they really are, of seeing things in their roundness and wholeness, and his very successes become failures.

— E. F. Schumacher

When we stop working with our hands, we cease to understand how the world really works.

— Clive Thompson

1. Where We've Been

Truth Is

They say that you have to go forward, that you can't go back no matter what. That the truth is, history never repeats itself. They chant mantras like "highest use best use," and "get to the bottom line," without a clue what either utterance might mean, what is being undercut and toppled. And never mind expensive brittle playtoys dreamt up in boardrooms, sure to be foisted upon us, they're also building alleys with no room to turn around.

But then there come moments like this one, that signal grand shifts in direction, that make some dead-end decades, even centuries, seem like they had never been. Moments spent outdoors wandering among dead leaves, when we find the path underfoot that convinces us someone or something knew where it was going. That we have been looking for all along.

Out of the Past

Even after he bought a Farmall, then a John Deere, Edwin still used his team of horses for all but the heaviest plowing, breaking new ground. He felt tractors seemed to pack the soil too much, and on his small fields decreased yields. It is also evident that he enjoyed the contemplative silences, the rhythm and feel of the work being done with his team. As a neighbor would say, he liked hearing the soil being turned. The days may have passed at a slower pace, and at a smaller scale, but he was able to feed and clothe and provide for a family of nine for many years on those 160 acres. So efficiency was not one of his sacrifices. And as for sustainability, the family is well into its fifth generation of stewardship on this land; two of Edwin's sons and two grandsons till those same acres today with undiminished abundance.

In the long run they might have been fortunate that the fields carved out of these rolling hills cannot be made larger. From the standpoint of Earl Butz's "Get big or get out" dictum in the 1970s meant to stampede farmers toward the consolidation and practices of agribusiness, they were stuck at a certain scale of effort and reward. But they never saw the small fields as a curse. And although the equipment has gotten larger, their dependence on crop rotations and manure as fertilizer and on a diverse and flexible enterprise have remained largely unchanged for the past hundred years.

One question occurs: given their traditions, why haven't Edwin's sons and grandsons continued using horsepower to work their fields? They might say they are of another generation, one that considers feeding work animals over the winter a pure waste, where fuel can just wait in the tank for spring startup. Maybe in the end they bought some of the advertising that has come at them lifelong in waves. Or given the traditions of a culture and a nation on the march toward some technically sophisticated future, maybe they could confidently assume they would never need to know what they considered the "old ways" of the generations before. After all, those ways entailed strenuous effort and occasional hardship, even if they also gave rise to a sense of common purpose and enjoyment of a fellowship in the work—what we can't help but call community—that is now lost to most of us.

Or they could say they never quite learned to work a team, that that was one of Edwin's special skills, one of his quiet definite ways they knew enough to stay clear of. So now driving their huge rigs up and down they may resemble agribusinessmen, though in their hearts and minds they are still walking in Edwin's confident footsteps, behind his mismatched team.

There are two other seeds to drop in this furrow. One is that Edwin's sons and grandsons operate a manure spreader that could handle the weight and volume of the largest Cadillac. Though Edwin might shake his head to see it, he would still recognize and approve its use. Why? In the early 1930's, before electric power was strung out to the region, Edwin lit every room in his house with an ingenious system of carbide lamps all rigged to a common tank. For most of his life he was viewed by his neighbors as technologically savvy, testing the latest solutions, restless with easy answers, always looking out ahead. So those horses— and the work they entailed and abetted—represent a considered choice.

A Good Start

Let us begin with the obvious, by asking what is farming. It is a livelihood of planting, tilling, tending and harvesting the bounty of the earth. A livelihood that calls upon a complicated set of skills maybe ten thousand years in the making. A livelihood that contrasts at a fundamental level with hunting, fishing and gathering, where those looking for food do little or nothing to sustain the sources of what they eat. Fishing, hunting, and digging roots can be hard work, or in the event of a windfall unexpectedly easy, but it is exploitation, a form of extraction, like oil drilling. Hunter-gatherers range in widening circles, and with no thought for tomorrow might well eat the last of the species.

Farming from the first has been a way to stay put. Which means that unlike nomads you can accumulate tools and books and arts and amusements—means alongside crops you grow a culture. Think of the Egyptians and their relationship to the Nile, the flooding of that river that fed a civilization on one spot for thousands of years. You could say that they didn't have to think about it, just feel that life-giving rhythm that ran deep as any heartbeat. Sure, they prayed, offered sacrifices. But then the river was dammed to run electric turbines, pumped for irrigation, its flooding timed, controlled, regulated. Perhaps it's just that we now pray to different gods. But there could come a day when the choice might arise between hydroelectric power and the lifegiving floods. As all dams are in time silted in, rendered useless, the Aswan dam will one day have to go, the valley return to the vagaries and rhythms of its annual flooding, that will again nurture farming on the Nile.

Historically, we harbor both urges, to husband resources in place, and to seize what we can and move on. Farming has more than once gained and forgotten the know-how that would over time let it linger and belong. It has been by turns greedy and wise. So the story of farming contains loops and detours. It has never been one endless furrow.

But there are some givens. All farming began as subsistence, as feeding oneself and one's family first and last. Apart from the oppressive bygone practices of serfdom and slave labor, farming only gradually touched the wider community beyond one's neighbors, offering occasional surpluses that would create an anonymous market. The essence of subsistence was diversification, where the effort spent on a variety of plants and animals meant a measure of safety against the catastrophic failure of any one. The notion of a cash crop may seem time-honored, particularly in the American South, but it was only recently that any farmer would think to designate himself as a specialist, as a wheat or pig or peanut farmer. Farming began everywhere with one or two crops, followed by an instinctive urge, prompted by self-preservation, to see what else might grow here. It took a while for the rewards of any specialization to outweigh the risks. It also took a while for any repeated, methodical set of tasks to seem appealing, alongside the diverse challenges of traditional farming, where for most of the year the work might change day to day.

First Farmers

Many centuries before the incursions of Europeans on this continent, Native American tribes developed their own farming crops and techniques. Their fields of maize, squash, beans, sunflowers, goosefoot (quinoa), amaranth and peppers allowed for permanent settlements, whether among the Northeastern Iroquois federation, the Mound Building peoples of the Mississippi and Ohio basins, or the pueblo cultures of the Southwest. Their three primary crops, known to the Iroquois as "the three sisters," were usually grown together in a symbiotic relationship. As cornstalks would reach up, the beans planted alongside would climb the stalks for support, while the squash fanned out underfoot as ground cover, blanketing weeds. Using this technique, along with an occasional fish buried in corn hills as fertilizer, the soil would remain viable for several decades, before the tribe would need to clear new ground. Dried maize and beans would keep relatively indefinitely, and allowed for the added stability of food reserves.

Though many of these tribes also hunted and fished, this was the seesaw path connecting hunter-gathering and farming around the world. From season to season their diet may have included wild meats, fish, roots and nuts, along with their staple crops. And all along there was innovation. It is likely that native farmers first discovered how to manipulate the genetics of corn, developing strains that could thrive from the tropics to New England. Northwest tribes along the Snake and Columbia rivers were well on the way to domesticating the Camas or Quamash plant for its edible bulbs. The pueblo peoples tamed and bred wild turkeys, and with the coming of Europeans would soon add sheepherding and weaving to their farming mix. Most of the plains tribes quickly became master horsemen and, particularly the Nez Perce, fine equine breeders and trainers.

This promising foothold in farming was largely erased by three hundred years of European settlement, as tribe after tribe was decimated by disease and war, and displaced by the incessant pressure to move out ahead of intruders. The Trail of Tears removed even accommodating native allies from the Carolinas—the so-called Five Civilized Tribes—to the west in the 1830s. Then, after the Civil War, the U.S. military was given the task of rounding up warlike nomadic tribes in the west, and forcibly converting them to farming. The policy was a painful and protracted failure, exacerbated by the fact that the reservation lands these tribes were confined to, as the leavings of white greed, were barren and marginal, mostly unsuited to tillage. Only the Navaho, Hopi and Zuni pueblo peoples of the southwest have been able to maintain an unbroken tradition of farming their ancestral lands.

Dirt Poor

For newcomers, subsistence proved a winning strategy, thanks to the continent's wealth of natural resources, and the example of the Native peoples' versatile lifestyle. And subsistence needn't always mean poor. Initially it can mean dirt poor, where your money is tied up in land, so you have little working capital. John Adams and his parents and grandparents offer a good example. They all farmed, but would look for something to do over the winter to make what they called hard money. You might take up shopkeeping, or make something at home like candles or woolens, start a cottage industry. Or learn surveying, as young George Washington and Crevecoeur did. Or like John Adams and Abe Lincoln, you might take up lawyering. Which is only one of the thousand ways rural people have tried to augment their income, and make ends meet.

Lead By Example

It might be worthwhile to contrast the lives of several of our most prominent early workers on the land. John Adams went into government from a career as an attorney which was financed and sustained from the first by the Adams family farm in Braintree, Massachusetts. At his father's death in 1761, John Adams inherited 40 acres and a house. As a boy he had resisted schooling, and told his father he loved everything about farm work, even slogging in the swamp. When he retired from the presidency he returned to a farm he had bought in nearby Peacefield with Abigail, to work its 40 acres, referring to himself with a chuckle as "Farmer John Stonyfield." At his death in 1826 his estate was worth $200,000.

By contrast, his intermittent friend and rival Thomas Jefferson owned a plantation worked by as many as 150 slaves at a time, in large fields of mostly tobacco, what today we would call a monocrop. Following the death of his father, Jefferson at fourteen inherited a 5,000-acre plantation and dozens of slaves. At the death of his father-in-law, he inherited another 5,000 acres. Yet by the time Jefferson died, Monticello had shrunk to 552 acres. He had a restless, fertile mind that reached deep into most of the arts and sciences, including agriculture. But he was also prodigal in his expenses, reckless in his investments and experiments. At his death his remaining estate, including its slaves, had to be sold to pay some $200,000 in debts.

A middle road might be found in the plantation efforts of George Washington. Though also a Virginia planter who worked 100 slaves, this founding father had a more diversified and practical operation, with some 200 acres of his 2,000 (6,500 at its height) under cultivation. He managed the land as five separate farms, kept accurate records, and shifted the main cash crop of Mount Vernon from tobacco to wheat, for which he built a flourmill, to move away from exporting grain to England, and feed a strong colonial market. By many he is considered the father of the American working mule, for his efforts to improve and promote the breed for work in hot climates. He also had a seasonal fishing operation that netted the spring run of shad on the Potomac, to be shipped smoked or salted in his own barrels, to British and Dutch colonies in the Caribbean. Besides a cooperage and a flourmill, the estate also built and operated a state-of-the-art distillery, and was profitably engaged in stock breeding, spinning and weaving.

Portable

Traveling midwest byways and backroads during his 1830s sojourn in America, Alexis de Tocqueville noted that subsistence farming along the frontier nearly always included three staples—corn, hogs, and whiskey. Hogs were an obvious choice, since in the backwoods saying it was often "root hog, or die"—hard times when hogs would have to forage underfoot for acorns, roots and nuts. The one crop that could be grown and eaten, fed or traded anywhere was corn, with hogs and whiskey as alternatives to make that corn portable—and potable.

The Farming Bug

Coming to more recent times, we might visit a pair of small farmers to whom we owe much. Luther Burbank (1849-1926) and George Washington Carver (1864-1943), were both known for their practical plant and crop innovations. Both have been viewed with skepticism by agricultural academics, dismissed as scientific amateurs since neither kept meticulous notes or conducted formal experiments. Both were too busy garnering practical results. Perhaps their backgrounds will suggest why.

The thirteenth of fifteen children raised on a Massachusetts farm, with an eighth grade education, Luther Burbank had the heart of a small farmer, buying his first 17 acre farm at 21 with a small inheritance following his father's death. There began his real studies; there he developed his first improved crop, the Burbank potato. With the $150 he was paid for it, he moved from Massachusetts to Santa Rosa, California, where he bought 4 acres of land, and widened his experiments to include citrus fruits and flowers, even cacti. He eventually bought 18 acres in Sebastopol, which he named Gold Ridge Farm. In all he developed or domesticated 262 fruits, 9 grains and grasses, 26 vegetables, and 91 ornamentals, many of which continue to be mainstays worldwide.

Born into slavery in Missouri, as a free youth George Washington Carver moved to Kansas to pursue his education. There he homesteaded on the prairie, plowed 17 acres of his claim by hand without the aid of animals, planted corn, rice, and garden produce, as well as fruit trees, forest trees and shrubbery. There he maintained a small conservatory of plants and flowers and a geological collection. There he also worked as a ranch hand and did odd jobs in town.

Though he earned fame as a teacher and researcher, Carver's farming skills were as much in demand to feed the students and staff of Tuskegee Institute through its early years, and support it with the sale of surplus crops, as any classroom or lab work he did. Recalling his own beginnings, he worked to break the economic dependency of Southern sharecroppers on cotton, and pioneered crop rotation to fight the soil depletion of that cash crop, focusing on peanuts, sweet potatoes, soybeans and peas. He preached a quiet, steady message of diversification and fertility. To build a foundation for peanuts as a crop, he offered 105 new peanut recipes and developed over 100 alternative uses for peanuts, ranging from cosmetics and plastics to dyes, paints and fuel.

Both men were eminently practical in their approaches and pursuits. Both consciously looked to empower the small farmer, and share the living tools for a more diverse and widely based success. Though they were childless, both were mentors to children, and influenced statesmen by their advice and example. Both were committed to spiritual and philosophical inquiry. Carver's accomplishments in poetry, painting and race relations also offer us a model for the well-rounded individual, capable of reaching out in new directions, while keeping a firm footing in what matters most.

The Straight Dirt

Good farming soil doesn't grow on trees—it grows beneath them in leaf mold, and in the cane brakes and lush growth of river bottoms and flood plains. It is a blanket of the living gone before. What we call dirt is the Middle English variant from the Old Norse, of *drit*—excrement—hinting at its connection with manure. Manure itself is an ancient euphemism, the Middle English word for cultivating soil, from the Vulgar Latin *manuoperare*, literally to work by hand. Rich crumbly dirt, loaded with decaying organic matter that will grow things, is not inexhaustable and should not be allowed to wash or blow away, not be worked when it is too dry, not be wantonly stripped or paved or built on, simply because of its location. It needs to be treated as a prime investment, nurtured over time by regular deposits and careful husbandry, since fertile soil is the bank that holds our future.

Dirt hasn't always been seen that way in this country. The original pattern of farming all along the frontier—except for rocky New England, but including most of the South—with few exceptions was exhaustive and exploitative. Fields would be cleared, corn or wheat or cotton planted until yields dropped low enough that fields would be left fallow, at which point the planter would move on. Tom Lincoln moved his lanky family five times in twenty-four years, from Elizabethtown, Kentucky to Little Pigeon Creek, Indiana, to Macon County, Illinois, and finally to Sangamon County, Illinois, chasing a decent yield for his labors. This pattern was followed by even large landowners, who would continue to break virgin ground on their holdings, then ultimately move west when they ran out of fertile soil. In certain areas the original topsoil might be more substantial, nine to twelve feet thick in parts of Kentucky, six to ten feet thick in the Willamette Valley and California's Central Valley, up to twenty feet thick in North Dakota, and in such places the lucky farmer could count on a longer tenure of plenty. But the hard lessons of soil conservation seldom rose to consciousness until after the frontier effectively closed in 1890.

But even without the allure of free land, and early farming's extravagant binges, change was slow coming. It wasn't until the manmade Dustbowl disaster of the 1930's blotted out the sun for days across the Great Plains, stripping a hundred million acres of topsoil off an area reaching from the Texas panhandle clear up into central Canada, that erosion created enough alarm to provoke changes in how things were done. Farmers adopted the two still common practices of contour plowing and strip cropping. And a third notion—leaving some marginal lands well enough alone.

In Continuous Touch

Farming is a continuum. From the first semi-intentional stick scratch in the dirt to the latest 8-wheel-drive diesel, we are all more or less connected to the mysteries. The science involved is both workaday and magical. Seeds go into the ground, something green and alive opens up. In season calving and lambing ring their changes. So farming remains for humans a vital source of meaning, purpose and value.

And we farm to do more than visit the mysteries; we crave to live in continuous touch with the sources of life. To work where and when we can, watch and wait for what comes, and let go the rest. Which can be both an art and a form of prayer. There is an essential modesty to diversified farming, a little this and that, which keeps us from the delusion some folks fall into, that approaching the sources of life and power, they come to think of themselves as the source. Farming partakes of a simple but potent thought—that just as we are omnivores, so no one thing should feed us every meal—that diversity and crop rotation do more than guard against failure. These practices connect us to sustainability, the long-term health of the planet, which is another way of saying that, beyond the alarm clock and the season's urgencies, they help us get up each day and carry on.

Sweet Water

When I was a boy the neighbors who helped us with haying always brought their own water—in glass and crockery jugs, left in shade at the end of the field, under wet burlap. When we would flop down for a rest, they would pass their water round, and take to sampling each other's, smacking their lips like connoisseurs. Newly dug wells would be judged alongside the older tried and true. All were hard water touched by local limestone, leavings of that vast ancient inland sea that didn't lather much in the wash, and came from hand-dug wells in a region no more than a few miles across. Yet there were differences. Some wore the faint bite of iron or other trace elements, and some were definitely sweet.

The Essential Imbalance

If we leap too quickly to the bushels per acre, and begin to count too heavily on the dollar equations, the cash conversions, we may lose sight of the fact that what we are investing in are the everyday sources of life, of sustenance. That must depend on an eye to the weather, count on seed corn in the crib, as well as an eye to the market. There is always an imbalance to farming, a world being born amid a world dying, somewhere under hail a crop beaten flat, a demand this moment for effort that may or may not be rewarded in time to sustain us. We only know for sure that there is always a crop and a living to make, somewhere out ahead.

Let's think for a moment about the illusion of improvements in technology, what is gained and lost. Running a tractor down endless rows without a kill-switch, if you fall asleep you could well lose your life. As a neighbor of mine did, in a steep ditch. And there was a Canadian farmer a few years back who fell off his tractor harrowing in a circle, where the harrow ran over him seven times before anyone could stop it.

But if you were to fall asleep working a team of horses in the field, the horses would sense your grip go slack and most likely slow to turn back and take a peek. Then if you were far-gone they'd take up the personal business of looking about for a mouthful. Unless they're not so well trained or content, in which case they might startle themselves with the specter of a blowing leaf, rear back and do something silly or reckless to wake you.

Bang For the Buck

There used to be a good living in farming. For a long time there was also a whole different set of terms for that success. Farmers doing well weren't called rich or wealthy—they were prosperous—literally hoping out ahead. It was called getting by, a decent living, not half bad. In their world of understatement, you might not be able to tell who'd had a good year, except by the fresh coat of paint on their buildings.

One Seed

But for a while now our culture has nurtured the delusion that every one of us could always be doing something more profitable. Could and should. Leverage our assets, get more bang for the existential buck. So we are encouraged to chant mantras like "other people's time, other people's money," and restlessly bestir ourselves, rejecting the work at hand while we scout out the next Main Chance. In effect gambling on our future. Yet when we consider the seduction of efficiencies and the supposed curse of inefficiencies, time in the furrow is sometimes all we save. And that economics can prove dubious, even ruinous, since soon enough we'll wake to find ourselves in some other rut of our own making. Say in front of the computer, combing through columns of figures trying to make something intangible and unreal somehow come out even.

Levitation

The Amish have long ignored such seductions, and calculate their practicalities to a fine fare-thee-well. Consider how they moved a building a few years back. Bolted pipes around all four sides at just the right height, then got neighbors by the scores and hundreds to help lift it off its foundation and together walk it to its new home. They didn't just calculate the cost of heavy equipment and professional riggers minus the cost of the steel pipe. They worked out how many hands and handholds they'd need for the weight. And the owner of the building wasn't the only recipient of largesse. The community profited, gained an enhanced sense of itself, of its power and identity. Everyone who helped that day learned how a community means shared effort, shared risks and rewards. Shared burdens made lighter. The antedote to isolation, the warmth to thaw the chilly cash nexus. So should one of them need a large job done, he would only need to plan the task for more hands, and spread the word. He would also harbor the urge to be a good neighbor well in advance of his need.

2. Where We Are

Where to Begin

The most burning and urgent farming question first: is it too late? Can we not turn back? And is it a question of will, of lock-step intractability, or of self-delusion ratified by the rut of the day-to-day? Once all the apparatus and economics—including government subsidies and market speculators—of monoculture are in place, can we not re-envision the land, work it again in smaller plots, with more variety, less reliance on petrochemical inputs and diesel-powered behemoth equipment? Can we not find ways to turn out cattle to graze on grass once more, without an uprising at the supermarket demanding cornfed fat-mottled beef? Are we unable to imagine any other ways of feeding each other and ourselves?

Or is it that we would then find the reins back in our own hands, and be faced again with the task we have largely surrendered to the grocery chains and big box stores and fast food look-alikes?

We've long been told the game is over. By 1991, less than 2% of Americans lived on farms. And of those who did farm that year, only 68% of farm managers and 14% of the farmworkers they hired lived on the land they worked. By 1993 the U.S. Census Bureau announced that it would no longer be counting the number of Americans who live on farms at all.

Though "farm" and "firm" are cognates, from the Latin "firmus," (firm, strong, stout; to be relied upon) capitalism coupled with farming has been an unhappy marriage, one that often seems to court catastrophe. Think of the falling produce prices and farm foreclosures that began in 1923, in the midst of the Roaring 20s bubble, a stagnation that prefigured the Stock Market Crash of 1929. Sensing our economy's and culture's addiction to growth we should at least think to ask, in what ways is bigger better? And what about our commodity futures market makes us more secure? The market economy as sole arbiter of viability has led to inhuman consequences. And with the United States' role in today's Global Economy attempting to outsource all but the profits, we grow increasingly deluded and confused. The farming label is borrowed to advertise enterprises having little or nothing to do with growing food on the land, for the residue of virtue in the name. Yet the values and skills learned in the effort of growing food gather in the minds and hands of those who do the work. There is more at stake than appears in just the plucking and eating.

But consider that fundamental, modest truth—that farmers eat what they plant. Often this includes chewing a little of the stem, leaves, seeds, and dirt. Lifting a little of what's grown to the mouth, to taste the strength and goodness of the life, sense where it comes from, what supports it. So crops are sniffed and tasted to see when they are ready for harvest. So many a farmer will deliberately taste and savor the soil itself.

The New Small Farming, in its values and ideals, is not just the old farming dressed anew. Decidedly not fresh lipstick on the pig. There are endless innovations, not just in equipment and practices, but in the training of draft animals and in the mindset of teamsters. There is a reliance on the sturdy substrata of traditional farming, on crop rotation and manure as fertilizer and on the traditional varieties of plants and animals. But then science and know-how have joined hands to find new methods of pruning, new tools and ways of working what to the unpracticed eye might appear to be the same old crops.

If we count what are dismissively referred to by the Department of Agriculture as "hobby farms," include urban and suburban gardens, and the swelling ranks of intensively cultivated inner city gardens on vacant lots, as well as those truck farms and gardens supplying a burgeoning list of new farmers' markets that have sprung up nation-wide in the past couple years, then we can imagine a potential food growing community numbering in the millions. The USDA reports that farmers' markets have grown nationwide from 1,750 in 1994 to 4,475 in 2007. In Philadelphia over 150 vacant lots have been turned into community gardens and parks. In a partnership with the city of Seattle, 78 urban P-Patches, amounting to about 25 acres, half of them owned by the organization, are tilled by thousands of neighborhood gardeners. In Washington D.C. an abandoned ballfield has become an urban community farm. In all three cities many tons of fresh produce are donated annually to local food banks. And consider where we are right now. In 2009, 43 million families planted gardens, up 20% from the previous year. So this is a swelling movement afoot, to know where our food is coming from, to keep meaningful contact with farmers, and share the work of feeding ourselves.

Perhaps the place to start is to examine how it has been done. What's wrong with the agribusiness model? It is after all efficient in its use of labor and inputs like fertilizer, insecticides and herbicides, isn't it? Or is it?

Not So Fast

The increased productivity in field crops in the past fifty years—the so-called Green Revolution—has come from three questionable agricultural innovations, until recently touted as little short of miraculous. The first is petrochemical: fertilizers, insecticides and herbicides that are natural gas and oil based. The second is irrigation: the widespread practice of buying up, pumping, channeling and subdividing the rights to ground water, with little accommodation or thought to the natural world. And the third is genetically engineered seed: interfaced with chemicals these marvels promise to deliver plants resistant to insects and tailored for local growing conditions, that shift the plant's energy from stalk and leaves to fruit. Such genetics have been carried to the point of engineering fruits and vegetables with qualities for improved picking and shipping, that make the most of soil conditions and rainfall in a certain locale.

How well do these inputs work? After all, it would be pointless and churlish to argue with success. The semidwarf wheat and rice strains developed since the early 50s are receptive to larger nitrogen inputs, their short sturdy stalks increasing yields by up to sixfold. Yet there is a downside. Nitrogen sources are nearly all petrochemical, unavailable in much of the third world, and unsustainable in the long run. And since 1950 pesticide use has grown to 2.5 million tons annually world-wide, while year-to-year crop loss due to pests has remained relatively constant. Which argues that pesticides have a limited value, and may have already reached the saturation point. The major difficulty is that over time these poisons select for pesticide resistance in the pest population, leading to a condition known as the "pesticide treadmill," in which pest resistance simply perpetuates the development of new pesticides, while the bugs always stay one step ahead. And consider just the human side effects: the World Health Organization in 1992 estimated that 3 million pesticide poisonings occur annually world-wide, causing 220,000 deaths.

When I was a boy, the best corn yields on the Indiana fields we worked in a regular four year rotation were about 80 bushels an acre, with 50-60 a fair average. Though admittedly we'd usually buy hybrid seed, our other practices were all traditional. We'd cultivate to keep down weeds and rotate crops to replenish the soil and stay ahead of pests, and once every four years spread manure and lime on the fields. Today much of the Midwest Corn Belt using the current panoply of chemicals, tools and seed expects yields of 160 to 190 bushels an acre, three times our old average. But these yields have proven elusive and erratic, with chemical equations that refuse to stay constant. The means and methods of agribusiness lock the farmer into a steady escalation of inputs to maintain that miraculous yield, ever more petro-chemical fertilizers and insecticides. Some oldtimers would say the farmer is bailing a leaky boat.

On a parallel track, increased productivity in livestock has come from three sources. The first is improved genetics, which enhances the growth rate and meat-to-carcass ratio of the slaughtered animal, though at an incalculable loss of genetic diversity and healthy breeding stock. Then there are growth hormones and antibiotics that ensure the health of the animal and speed the race to market weight, though at an untold risk to consumers and to herd animals' longevity. And finally, new penning and feeding practices, which have all but replaced the "free range" lives of cattle, hogs, turkeys, chickens, sheep and other farm animals with a prison-like confinement, that also threatens the public with its massive downstream pollutants, and reliance on dubious processed protein inputs.

Along with these increases in productivity have come the so-called "economies of scale," as the average farm size over the past forty years has more than doubled, from about 200 to over 500 acres. The argument is that equipment is more efficiently employed when working large monocrops round the clock. The metaphor is industrial, actually "machining" the land in ways analogous to strip-mining and clearcutting. Often with similar environmental results. It is chemically assisted fieldwork that attempts to deconstruct and reassemble nature. Some critics point out that the paperwork gains in productivity have mostly come while ignoring the actual costs to the environment and to one's neighbors. In the short run, agribusiness may make money, but only by emptying out the farming community, and abandoning any notion of sustainability. Saying the hell with the land and the rest of us.

As for sustainability, these large-scale monocrop practices have resulted in massive losses of topsoil—in Palouse wheat country, for example, seven to fourteen tons per acre per year. As a further side effect, the use of chemical fertilizers often masks soil erosion, so that topsoil loss can go unnoticed until it is too late. And these high concentrations of nitrates, phosphates, and feedlot manures in streams cause algae blooms and loss of native plants, fish, and other wildlife.

Whether due to greed, indifference, or a dazzling job of salesmanship, the agribusinessman has found himself on an accelerating treadmill, maximizing production to meet fixed debts. Bigger equipment has led him to plant more acres, while at the same time commodity prices have been driven down, threatening to drop out from under him. While the alternative of diversified farming is dismissed as a romantic bygone ideal, the operator is encouraged to grow more corn to get more money to buy more land to grow more corn—in what, considering fertility and other environmental costs, must seem an endless downward spiral.

For a while there, from the 60s to the 90s, as the mechanics and chemistry, business practices and politics of agribusiness fell into lockstep, it did seem like survival of the fittest, where the few farmers who bulked up and survived were truly the best of the best. After all, something in them was capable of meeting the rapidly changing demands of government and technology, threading an impossible maze of falling commodity prices, government regulations, and rising mountains of debt. Besides which, everyone knew the old ways were defunct. The message to farmers on every side was face it, this is the way forward, the only farming game left.

So at heart the proponents of agribusiness nurture the delusion that their ways represent the future. Yet the facts are stubborn. Farming employs over a third of the world's population, yet agricultural production today accounts for less than five percent of the gross world product. Which suggests how the work and its output are devalued in the marketplace, with the lion's share accruing not to growers, but to corporate processors, transporters, distributors and retailers, porducers of hybrid seeds, fertilizers and pesticides, equipment and fuel. As a reality check, we need to remember that despite synthetic fertilizer production and use in the United States, most farmers worldwide still

employ manure and compost. And most farmers worldwide still work
their fields with animal and human power.

Corn Jabber

To work what he calls the corn jabber, you hold the wood handles
apart, stick the iron-tipped duckbill into loose dirt, then push the
handles together, which opens up a little slit. At the same time from
the hopper fed by its little ratcheting wheel underneath down will drop
a kernel landing just about right, after which you pull apart and lift the
whole thing up. Does practically everything but kick dirt in the hole to
finish it. Their oldest, simplest planter from century before last, this is
what Edwin's sons still use wherever ground moles get after the seed
corn, hoping to save themselves a little bending down, fingering each
new life in the dirt. No matter how big the field, worked double-tired
four-wheel drive diesel everything, still they climb down to plant on
foot by hand, wherever nothing comes up.

Myths About Large Scale Farming

The truth is that the highest yields on the planet have always come and
are still attained in small plots, in rice paddies and hand-tilled home
gardens. It's the old story of personal motivation, concentrated effort
and attentiveness, reaching back to feudal serfs in Europe, to the slaves'
and sharecroppers' gardens in the American South, and to the peasant
terraces of China, Korea and Japan. Private garden patches grudgingly
permitted under Soviet regimes fed the nation through times of severe
privation. For eighty years the people of the old USSR fed themselves
despite the active discouragement of local Party officials, and brutal
programs to collectivize and farm the land in common. It was no
mistake that during WWII Roosevelt called on all patriotic citizens in
the United States to plant Victory gardens. Those gardens
strengthened families and fed local communities, while freeing up
resources for the armies overseas.

Voicing a wisdom still alive among our own traditional farmers, the poet, essayist, novelist and horse farmer Wendell Berry has long advocated that we should work few enough acres that we get to know and love them. So that we can watch over and care for everything that happens there. With the implication that over time we might learn from the land, be nurtured and improved even as the land improves, and in the practice come to know ourselves.

We also shouldn't be misled by arbitrary pigeon-holing. For instance, gardening can be seen as another form of subsistence farming, in that both grow food for oneself and one's family, on an intimate and efficient—and aesthetically pleasing—scale. Yet in the media we are often sold a class distinction about gardening, that it is a leisurely pastime for the well-to-do. But the word garden at its root just means an enclosure, a plot with a fence or wall surrounding it, where close attention can be paid, and in time rewarded. The view of gardening as an aesthetic pursuit might baffle most of the farmers I knew as a boy. Yet alongside lovingly tended fruits and vegetables, their gardens were always lush with flowers.

So what does scale do for farming? As practiced at the contemporary industrialized extreme, vast monocrop fields justify the status quo, which is a massive ongoing investment in equipment and fuel, irrigation, petro-fertilizers, pesticides, herbicides and bio-engineered seed. Never mind the equipment operator's endless hours at the controls, this mechanized, self-consciously "efficient" effort is aimed at one target—a maximum yield for a minimum price. All other values—taste, nutrition, sustainability, quality of life for the farmer—have been sacrificed to this so-called bottom line. In a small patch weeds can be pulled or hoed, not indiscriminately poisoned. In a small patch plants can be thinned as they come up, vegetables picked as they ripen.

Within Earshot

Come spring there is always that one hen hiding her clutch in the weeds or up in the loft somewhere, trying to lay low a few days, stay invisible, just long enough to hatch out a brood. Which is more than a game, we both know it. After I have found and raided in turn all the others, I stand still in the yard a few minutes, listening. Sometimes I am distracted or rushed. I don't always find her, but often enough. If I can just wait her out, the proud hen can't help boasting a little, her quickened clucking down low to the ground broadcasting her secret.

Race to the Bottom

Looking for where things have gone wrong with the ideas and methods of farming, maybe we need to widen our search. In his 1973 book Small is Beautiful: Economics as if People Mattered, the visionary thinker E. F. Schumacher brilliantly exposed traditional economics as in effect a pseudo-science, citing how it ignores the limits of growth inherent in the carrying capacity of the earth, and how it recognizes no distinctions between sustainable and non-sustainable goods in its bottom-line approach. He pointed out how disastrous it is to equate primary goods, both renewable and non-renewable, with secondary goods, including manufactured goods and services, which depend for their existence on those primary goods. A dollar's worth of wheat is different in kind and ultimately more valuable to us than a dollar's worth of oil, and both are preconditions for a dollar's worth of nails or a dollar's worth of computing or hair styling. The cash nexus levels and dismisses such distinctions, in its race to the bottom.

Then there is the nature of the marketplace, which Schumacher saw as a ruinous jostling of bargain hunters, whose search for the lowest price at the cost of all other values has led to what he called "the institutionalization of individualism and non-responsibility."

At the same time we have to admit how effective the salesmanship of agribusiness has been. Who knew that seducing the farmer into investing in outsized equipment and synthetic inputs was an impressive step toward effectively erasing that lifestyle and craft altogether?

We shouldn't be surprised at how agribusiness technology and techniques affect the farmer. In the post-industrial economics of wealthy nations, labor may still be given lip-service as the source of all wealth, of added value, but in practice labor has become an all but negligible part of the equation, an inconvenience or nuisance that management circumvents by intensifying its use of technology. What's left for workers to do is often mindless and repetitive tending of machinery, or meaningful tasks broken down into their meaningless constituent parts, mere muscle twitches. As a clear example, currently the building of a new car, with all its separate systems that converge on an assembly line, takes less than ten hours of salaried worker's time from start to finish. All the rest is tooling and robotics, neither of which need health care or a day off. Management's complaints about labor demands are largely a distraction from more substantial issues— the malaise and disenchantment of a buying public over what they're being offered.

There have been experiments to temper these industrial processes that would reduce the worker to an automaton. A Scandinavian car manufacturer at one plant tried having each vehicle assembled by a team, a group that would move along the assembly line with "their" car, doing each task in turn. The findings of that experiment are no surprise: much greater satisfaction among workers, and improved quality of cars—though sadly this simulation of a handcrafted product is considered too costly to attempt on a larger scale.

Most of the recent technological innovations in agriculture have consistently favored larger farms, with the overall effect of substituting machinery and chemicals for labor. From 1950 to 1990 labor costs fell from 38% of total U.S. farm input to about 13%. Over the same time, purchased inputs, including machinery, fuel, fertilizer, chemicals and seed, have increased from 45% to 62%. Labor as a percentage of the cost of growing food has been consistently dwindling.

So it's hard to tell whether or not we should pity the ag-biz operator for those stultifying long hours in his air-conditioned cab. There is another labor element in current agribusiness that is scarcely mentioned, though it's omnipresent—the migrant worker.

Our current farm labor situation, with its hypocritical reliance on mostly low-paid, undocumented migrant workers from our Central American neighbors, is the logical effect of having destroyed an existing network of small farmers who once cooperated to help work and harvest one another's crops. The reliance is hypocritical because of the phony implication that these migrants are taking jobs away from American citizens who, without a stake in the game, would generally refuse to work so hard for so low a wage. And while there is no complaint about the quality of work being done, there is also no suggestion that workers deserve higher pay.

In a perverse extension of present large-scale practices, some California growers have recently concluded that the migrant labor problem in the U.S. is insoluble, and are chasing that farm labor back over the border. They have leased cropland in Mexico, and are trucking their equipment, seeds and chemicals south. There they displace Mexican farmers, tying up large tracts of land to grow crops for American tables, exporting the petrochemical agribusiness model intact. In this epitome of unabashed economic imperialism, all that is required of Mexico is cheap land and cheap labor.

Scales

Charlie was a big Swede, 66 when I knew him. Not fat, just big all over. Six foot five, maybe two-ninety, three hundred. Savvy and prosperous, seemed to be cut out for farming. One day combining wheat he and I filled sacks to toss on the wagon for feed, then unload back at the barn. On the scales each had to go sixty pounds minimum. Because it was fixing to rain, I took two in each hand the first trip, then after that breathless stagger backed off, content to sling one on each shoulder. For every trip of mine Charlie took up and shouldered three sacks in each immense paw. When I mentioned the godawful weight, he just spat, snuck a grin that faded as quick as it flared. Shook his head, looked off and muttered "I ain't the man I oncet was."

The Bigger They Come

As a corollary to the issue of scale, we might also consider the notion currently being aired, of some banks and corporations being "too big to fail." As if there could be an apotheosis of scale, a kind of canonization, where one is elevated to economic sainthood, considered part of the cultural bedrock. But in the marketplace there is no such thing. It may take a little longer for an AIG or GM to hit the ground, but if it cannot manage to offer a service or create a product people will buy, cannot turn a profit within a finite number of quarters, there are no pockets deep enough to sustain such a bailout. We smaller folk who must live and die by the marketplace are innoculated against such foolishness. Responding to a question about a depressed area where unemployment was twice the national average, Ronald Reagan blithely told reporters that's the wonderful thing about America, that everyone is free to move.

And there is a counter to the "too big to fail" notion in the David and Goliath story, where the bigger they come the harder they do fall. There are upper limits to scale, size and speed. We might recall that the Concorde jet was retired in 2003, and not replaced. Or consider a traditional manufacturing example, in the typefounding industry, where by 1900 85% of all printing type sold in the United States was manufactured by an early "trust," the American Type Foundry, who in the previous dozen years had bought up all its competitors. In a pattern of behavior mirrored in meatpacking, oil and other trusts, for most of the 20th century their business stagnated. Lacking competition there was no impetus for innovation. When ATF abruptly declared bankruptcy in 1991, the assets that were worth $70 million in 1900 dollars were sold or scrapped for $7 million. Needless to say there is a continuing demand for new type in fine printing, that is only partially served by several small competitors who grew up overlooked between the cracks of the corporate sidewalk.

To sense the vulnerability of the commodified big-box food world, we have only to consider the example of beer. Over the past forty years the major American breweries have been busy buying each other out, growing larger and more homogeneous till their products were nearly indistinguishable, with only three giants left jockeying for market share. But then all but unnoticed a whole new world of microbreweries sprang up underfoot, whose brews were instantly seized upon by those who like beer. The model caused consternation in boardrooms. Here were people willing to pay two or three times the going rate for a glass of beer, preferring a brew that was cloudy and tangy, fullbodied and flavorful—everything their products were not. The market quickly reached a tipping point, that had the big three scrambling to buy out, co-opt or imitate these upstart microbrews—which have in turn become a strong incentive for small farmers to raise heritage strains of barley and hops.

We need but imagine the failure of a major agribiz player, a big-box supplier in the food chain that has been vertically integrated, that owns or controls a vast monocrop acreage, plus all the picking, sorting, processing, warehousing, trucking and packaging on the way to the consumer. Say a Safeway or Kroger or Wal-Mart—or in Europe a Carrefour or Tesco. The repercussions of such a failure might stagger and astonish us. If the network of one of these big players were to evaporate, the fleet of trucks and mountain of bins get sold at auction, all those functions and jobs would disappear overnight, as if they had never been. And they would not be immediately replaced. A network of competing and cooperating small businesses could not be crushed by any such collapse; you can imagine individuals responding within hours to the increased demand, and one pickup load at a time taking up the slack.

At the Heart of the Issue of Scale

Let's return for a minute to the Founding Fathers, the framers of the Constitution. During the Revolution, when 98% of the colonists were subsistence farmers, the Founding Fathers knew they had to appeal to the strength, independence, and endurance of those who worked the land. These were the people who would sustain not only themselves but also the shopkeepers and legislators and armies in the field. Used to husbanding resources, they could wait out the British, when invaded fade into the back country and borderlands. They could hold on forever, biding their time till the moment to strike.

Among the changes these bewigged, eminently rational men could not have foreseen was the rise of corporations in the second half of the 19th century. These entities are by definition non-persons accorded most of the benefits of personhood while carefully limiting their liabilities. With an abiding concern for redressing the grievances of citizens at the hands of governments, and with a healthy skepticism about the size and power of government itself, the Founding Fathers would surely have had something to say about how ordinary citizens suffer at the hands of corporations, and how human rights might pale in comparison to those of ostensibly immortal and nonmoral non-persons. How humans in the flesh run a poor second to these paper entities.

What might they have done? Windfall profits aside, the founders might have forced corporations to declare permanent allegiance to one country or another, and meet some minimum requirements of citizenship. They might have imposed a legal lifespan on a Colonel Sanders, Jack in the Box or Walt Disney. At the very least, it would not be hard to imagine them calling an IBM or General Motors or ADM to the bar once a generation to review its behavior as a citizen, with the court empowered to disband miscreants. The shrewd observer may respond that such a company's officers would just incorporate under a new name in a neighboring jurisdiction. Yet the notion of punishing the actual owners and major beneficiaries of corporations is a fairly recent idea, which only began with the trust-busting efforts of Teddy Roosevelt in response to the scandals and excesses of his era. Recent experience with Enron, Halliburton and other high-flyers shows that the more light thrown on executives and their accounting firms the better. The more dubious the shell game, the more likely it's played in the dark.

The Founding Fathers were already familiar with lobbying and lobbyists, though that force has been elevated, from a knot of shabby petitioners lingering hat-in-hand in the shadowy halls of power, to a power itself by corporations and their economics of unfettered growth.

Lobbists purchase, peddle and swap influence, to affect the rules of the game as they are written and enforced, often to the point of helping draft new rules and exercising veto power. Thus they are a force for corruption, and clearly no friend of the small farmer. It is fundamental to our system that office-holders elected by the people are chosen to represent the public at large. The lobbying purchased by big money gives that money an undue voice and an unwarranted vote.

At bottom, it would seem that the deeper the pockets—or the deeper the subterranean pools of credit—the more successfully one can compete. Agriculture Secretary Earl Butz's "Get big or get out" mantra from the 70's was aggressive, offensive and dismissive. But he was merely the obnoxious mouthpiece for a deeply ingrained philosophy current in academic and economic circles, that small farming is the problem, for which agribusiness is the solution. Subsistence farming— what economists term smallholder farming— still accounts for most of the food grown outside the major industrialized nations worldwide. Yet many agricultural theorists assume that smallholder farming is just a passing phase on the way to industrialized agriculture. A way station—inefficient and temporary. That somehow the small farmer will grow up, get away from all that physical work and start to use fossil fuels and the panoply of industrial inputs to escalate his whole venture. But given the carrying capacity of the planet and its finite resources, this is impossible.

In a Pinch

We needn't look far afield for our own subsistence lessons—and to locate the curse of the monocrop. In 1936 the writer James Agee and photographer Walker Evans spent six weeks living with three sharecropping families in backcountry Alabama, an experience that resulted in their epic account, Let Us Now Praise Famous Men. One family of four they met lived on a yearly cash income of $18—which in season required everyone in the field picking cotton dawn to dark. All their other material needs were improvised—or met by the garden patch.

The Loan of a Baler

Practicing an admirable independence, small farmers have always relied on subsistence and diversity to get them through hard times. If prices dropped to nothing, they could always eat or feed what they grew, hunker down and wait out the blow. Secretary Butz boasted that modern large-scale farmers were becoming model consumers, buying as much of their food and other necessities as any city family. With the increases in scale, demanding ever larger credit outlays for equipment, fuel, seed and chemicals, food production has become practically an adjunct to banking. The agribusinessman has to have a steadier cash flow and a more predictable profit margin, because he has many more and larger partners to share the proceeds with. And relying on credit, accepting a minimum share of the proceeds, he is always the most vulnerable one in the partnership.

Farmers used to rely on credit of a different kind—community—to help them in a pinch. But those communities have been gutted as smaller farmers have been squeezed out, with disastrous effects. The landscape has been emptied, the few big operators left increasingly isolated. The view of a society as an aggregate of shoppers and consumers—and competitors—does not allow for many of their positive effects on one another. Doesn't allow for the loan of a baler in season, the sharing of a hay surplus with a neighbor in need.

The vertical integration of farming into agribusiness contains flaws that have for a long time been all but invisible. Controlling markets, fixing prices and focusing on one vast monocrop may help maximize profit, but that profit can come at the cost of autonomy, of the sense of one's independence. Owning our choices for better and worse, accepting our lumps and rewards can be bracing, and one of the basic drives that keep us moving through a life. There is a debilitating effect on the quality of all efforts where small independent players just want to get big enough to be bought up by the few big players, who reaching up are less connected to the ground from which we all spring. No manager of a fast-food restaurant, one of a chain of hundreds or thousands of outlets, can have much sense of being the captain of the ship, the master of her fate, when compared to the owner-operator of even the humblest lunch counter. With farming this is true in spades. For something in all of us, any boss can't help but be a bad boss.

Farming the Government

No citizen in the street enjoys thinking about what an ethical quagmire his Federal Income Taxes have become—what a sorry game. Any more than farmers care to reckon how much their survival over the past fifty years has depended on that perennial game of set-asides and subsidies. So often being propped up and waltzed around the market by government agents, essentially paid not to farm. Rewarded for how well they can fill in forms, profit from current hot trends in conservation, and play along. In effect, farming the government. But this is motivation from outside and above, a gussied-up array of carrots and sticks. And playing along we can forget to ask what it is we do poorly or well. Whether the sole determinant of our identity should be a government-manipulated market where lobbyists provide the invisible hand.

It may seem counterintuitive that federal farm subsidies should more than triple in recent years—from $10 billion in the 1980s to $31 billion in 2001—while the number of farms and farmers continues to dwindle. As a voting block you'd think they might run out of clout. But the large-scale farmer functions as window-dressing for others who make more out of the venture than he or she does. Agribusiness is joined in its lobbying efforts by oil companies, heavy equipment manufacturers and dealers, genetic engineering firms, chemical companies, banking giants, supermarket chains, fast food producers, and on and on. Those who stand to gain most from unfettered corporate farming. And not surprisingly, agribusiness remains the one major polluter that has not been held to account, in large part due to the mythic appeal to rural independence that the iconic "family farm" still represents—and to the lobbying of its powerful allies.

In this situation, having a subsistence element to our work can provide grounding, as can feeding our neighbors and friends. No matter what the larger market says our produce is worth, it has intrinsic value to family and community precisely because of where it comes from, the realized effort and skills it represents, the sure knowledge of its place in the scheme of things.

Farming as Identity

I drove city bus for a couple of years. I noticed then that although bus driving was a necessary and vital function in the life of a major city, none of the drivers was comfortable being identified as such. It may seem strange that anyone smart and skilled and emotionally stable enough to do this demanding job would feel demeaned by being called a bus driver. But as I negotiated the sprawling transit system, any time I met another of my thousand colleagues, I would find that driver quickly identifying himself or herself by some avocation. "I got into Cajun cooking and now I'm writing a cookbook." Or "I bought a string of rental apartments that I'm replumbing and rewiring." I met stone masons and master gardeners and inventors and concert musicians and on and on. Maybe even a secret farmer or two. Bus driving was how they made a living, but was almost never how they saw themselves, never what they'd have wanted carved on their tombstones.

The same was mostly true with the farmers I knew and worked alongside as a boy, though with one crucial difference. Often they did have other jobs, other skills, powerful interests—and were self-effacing. But to anyone who showed the slightest interest in their doings, how they arrived at their harvests and all this came to be, there was a quiet and substantial pride. In that old expression, they always had something to show for their labors. In fat years or lean, they could point to a grape arbor in bloom, a barn packed with hay to the rafters, a hillside full of healthy, contented cows. The results might seem plain, though the skills and plans and identity that informed the enterprise were neatly tucked away.

Perhaps any culture that would denigrate farming deserves what it gets, which in the end is starvation. There is a fundamental waste involved in making the farmer feel like he needs to prove himself. The world knows there is more to his calling, that he's not just a throwback with no clue how to get ahead.

Take This Hammer

It is curious that industrialization, which first found ways to replace men in their tasks, almost immediately stimulated the impulse in workers to imitate and compete with the machines they stood alongside. The legend of John Henry's contest with the steam drill is an archetype of the measuring that took place, of the strains and ultimate despair of that effort. Of course anthropomorphic early machines had arms and reciprocating motions, whispered and hummed their own mindless tunes, seemed to crave constant tending and adjustment. Coddling the great iron baby was an admired and even envied role.

So there is always a certain pleasure to working the controls, climbing atop the great beast, combining or plowing from on high in air-conditioned comfort, though the operator is isolated, both from fellow workers and from the land and its crop. Held at arm's length from a physical sense of what the work entails. The man or woman in the driver's seat can't help but feel like less of a farmer, more of a machinist than a nurturer.

Right Sized

Sometime in the 30s Edwin and a handful of neighbors went in together to buy a threshing machine. It turned out to be about the right sized group for threshing—willing hands taking turns mowing, raking, binding and stacking, hauling shocks to the thresher, feeding them in the top, tending to the needs of the machine. Filling and tying off and muscling the sacks of grain away under cover. Forking around the bright mounds of straw piling up. Every hand engaged, going around to all the farms in turn, to the next field to ripen. After a few years of this, the dance came out so perfectly they probably started to stagger their planting a little. The threshing of wheat, oats, barley and rye supplied these families a social scene, a seasonal excuse for day-long get-togethers, a parade of parties spaced over a stretch of several weeks, each gathered around a noon meal. You might imagine these working occasions with all the kids underfoot, the old folks sitting close by, as the largest party that could have been thrown on any one place. And with onlookers, reminiscing or newly awakened to the whirl, you might well envy the workers and savor the work all around.

Of course Edwin's was the largest spread, like the others with plank tables on sawhorses set out on the lawn in the shade. With three kinds of meat, green beans and baked beans, hot and cold potato salads, mashed potatoes and gravy, four exotic casseroles, fresh bread and cake, gallons of coffee and lemonade, iced tea and a mountain of pies. Each family strove to keep its end up, the women rising in the middle of the night to get the baking done.

Then several of the most prosperous farmers in the 60s bought combines, and this work that had relied on the same communal equipment for thirty years, that with variations reached back at least to the Civil War, vanished overnight as if it had never been.

Feel at the Controls

We might consider the multiplication of the human effect to help us gauge in the end how much is enough. If we could create a world where only one man had unlimited productive powers, who would want to live there if one were not that man? Might as well inhabit the world of a comic-book superhero, as one of Gotham's myriad anonymous victims. Then too we might consider the lessons of servo-mechanisms, the "feel" that has to be engineered and simulated for the operator, in the interest of safety. We are after all physical creatures that need to walk and lift things and work with our hands. So the measure of when any mechanism may be getting too big, is when you need servo-mechanisms to supply feedback to the controls. Think of early power steering on large cars and ships and aircraft, that could represent a danger, since the wheel could be effortlessly turned lock to lock. Now such controls are designed to let the pilot or driver know through his muscles, recreating that crucial measure of what his moves count for, when it takes more effort to make a sharper turn, or stop more suddenly.

Farming in its many cultures over the past ten thousand years has always come with its own measures, its own scales. The upper limit of the land a man could own might be how much he could walk around in a day, or with tillable land how much he could plow from sunup to sundown. Measures of how much the individual could accumulate, and how much accrued to his community. As the Amish practices remind us, a barn or house may not contain the impulse to be so grand if it is designed to be raised by one's neighbors in a day, by neighbors who might measure its refinements and capacities alongside their own.

How Much is Plenty

As workers we continually take the measure of the work with our hands and muscles, and even at the small end find a need for appropriate scale. The handle wants to fill the hand. Notice how some computers come equipped with "chiclet" keyboards too small to be typed on. Likewise the tiny gas and clutch and brake pedals on some cars.

The ultimate in modest scale, forty acres and a mule—the dream of every freed slave in the South—seems little enough to ask of a brave new world that holds such promise. Until the 1950's, a man could make a marginal living sharecropping forty acres, though on the traditional share of one third he had little likelihood of sustained success, what we'd call getting ahead. If he owned his mule and tools, though, success would be more likely, sharecropping on half-shares. But if he owned the land too? Forty acres might begin to seem like plenty, if it didn't all have to be planted in a cash crop spent to buy staples at the company store that owned the gin and scales.

So when and where do we begin to draw the line? And how much is enough?

One Seed

3. Where We're Going

Eatery

There's been a lot of talk about food lately, about nutrition and food safety, and with the weight gains of Americans, there is evidently more interest in eating than ever. At the same time cooking at home has been on the decline. We speak without irony of buying "comfort foods." There is a deep disconnect in the spectacle of folks who hate to cook but love to eat, a schizophrenic malady traceable to the infantile urge for immediate self-gratification. Or is it simply a surrender to advertising, that seduction perfected to fan our desire?

Even among grownups, in speaking of food as a processed or distributed or packaged entity, maybe we're getting the cart before the horse—or skipping the horse altogether. Just as folks can be considered as consumers—as shoppers and bargainers in the marketplace, as momentary users and discarders, as miniature whirlwinds of desire, gratification and remorse—so considering food without regard for its ultimate sources can seem limiting, insulting, even stultifying. Such food discussions can miss the point entirely. Food needs to be considered in the context of its sources in farming, where various life forms in growing appropriately support and feed one another, with the ultimate end of winding up at our tables, where we feed ourselves.

Speaking of food without speaking of farming assumes a foregone conclusion, that the industrial solutions of agribusiness are frozen in place, fixed in time. Assumes that the sustainability argument is over, and that the way ahead is self-evident. Just open your eyes, look around. The notion of consumerism may have begun as an empowerment in the hands of Ralph Nader, a way to push back against some indifferent giants, but it has long been co-opted into a sanitized whine about quality control, that Verlyn Klinkenborg labels "a willingness to be bribed into ignorance" when it comes to questions of farming and land use.

Of course we can be considered as consumers, divorced from knowing how to check the oil in our cars or boil and mash a potato, just as food can be seen as a commodity laid out before us, in a shopping context where presentation is everything. We can ignore the implied insult as we study so-called "fool-proof" instructions. But we need to understand how our food has been commodified, how we have been distracted by claims of convenience and "value added," tricked into indulging in food substitutes that are mostly processing and advertising—in other words, fluff and hot air. It may seem easier and more sanitized to view each prepackaged meal in purely economic terms, but that is what agribusiness wishes for us, a reduction of the human to the simpler plane of shopping and its illusion of limitless choices.

Although the basis of our form of government—the vote—is informed choice, it seems that reliable information is all too rare, a jealously guarded commodity. Given the scale of agribusiness operations, and the remoteness at which we get to consider problems of health and

One Seed

food safety, it's hard to tell whether anyone at all is minding the store. Perhaps scientific experts on the government payroll should be telling us unequivocally whether the cattle brain matter and other cow parts being fed to chickens, that in "feather meal" and chicken litter then legally find their way back into cattle feed, is a safe and sane practice, or just a way of evading the 1997 ban on this source of Mad Cow Disease. Failing that surety, we might well be skeptical, tell each other it doesn't make sense, and choose to rely on someone we know who says he would never let his cows eat anything but grass and hay.

There is that moment in the classic movie <u>Coolhand Luke</u>, where the Captain tells Luke he's got to get his mind right. Which on a chain gang means surrender to the powers that be—play along and keep your mouth shut. We've been told by the big economic players—and their spokesmen in government—that the new efficiencies and scale of things are all scientific, and healthy, and in our best interest. That we need to get our minds right about our place in the new scheme of things. Mow the lawn and pop dinner in the microwave and shut up.

Yet sometimes the mask slips. The microphone is left on and Enron managers are overheard joking about gouging widows and orphans, in a fictitious scheme to drive up the price of electric power across the American West. Bernie Madoff admits he has consumed $65 billion of investors' money in a Ponzi scheme without a shred of remorse.

Down the Gun Barrel

The future of farming faces a stark set of choices: given the current agribusiness model, there is an impending collision between population growth and food sources worldwide. Notice how the recent rush to embrace corn-based ethanol in the United States caused spikes in the world markets for all staple crops. We have seen how interconnected these larger players are, and how ill-considered are some of their choices. In the boardroom fear and greed compete with wishful thinking. And we have seen how little they care for the folks who need to eat and get to work, the so-called consumers, reduced on the ledger sheet to isolated bellies, gas tanks and wallets.

A parallel looming crisis concerns food safety. Never mind that the Pure Food and Drug inspectors in the United States are barely able to keep up with their current job of protecting the public. Given the scale of domestic suppliers and their assertions of self-policing, only a small percentage of products are independently inspected by government agents anyhow. Much of the federal responsibility has been passed on to individual states, where it is often ill-informed, as well as underfunded. The flood of agricultural products from abroad, both raw and processed, presents nightmare scenarios of both thoughtless contamination and deliberate sabotage. And these products go largely uninspected. The first sign of a problem appears at the "big box" outlets and fast food chains, at school lunchrooms and nursing homes, as random victims are stricken, and product recalls and warnings issue forth.

Need to Know

Why has buying food become such a crapshoot? Even with packaging and labeling laws, we are still largely kept in the dark about the sources and contents of what we eat. It may be sad to say, but the ignorance of urban and suburban food consumers is largely a function of accepting a superficial role in the cash nexus, divorced from knowledge of the land and its work. We have gained a certain sanitized level of information about our food held at arm's length, that in the case of processed and packaged items has recently come to include a breakdown for quantifiable contents such as calories, fats, sugar, starch, salt and vitamins. All wearing a respectable veneer of science—but nowhere near complete.

We also need to be familiar with the fundamentals of each crop, its appearance and uses. For instance, that potatoes should come to us dirty, and be washed as we cook them. That scrubbing them for market might make for a slick presentation, but hastens the process of decay. The same is true of those mechanically abraded little snacking carrots. We are left to imagine how in a Manhattan boardroom someone is calculating the number of pounds of potatoes or carrots on average that should rot in each kitchen, to increase corporate profits.

And how both distributors and retail supermarket managers calculate the volume of water to be sprayed over produce on display, to be sold by the pound with the product.

The cases of E-coli and salmonella present other less visible dangers, for which the eating public continues to be woefully ill-informed. And short of recalls and warnings to throw out food that might be contaminated, the public is largely left to fend for itself.

Closer to Nature

One buzzword that signals our growing concern with food safety and quality is "organic." But exactly what does that mean? And what should we be thinking and doing about it?

Organic commonly refers to plants grown using manure and vegetable compost, without synthetic fertilizers, herbicides or pesticides, and animals raised free of chemical injections and additives, such as hormones or antibiotics. The goal is transparent—healthy natural foods, feeding a life that is closer to nature.

But where is that? In nature the top layer of soil is inhabited by myriad life forms, from worms and grubs down to bacteria and enzymes. Three billion organisms in each gram of topsoil—which might seem either the epitome of cooperation or of competition, depending which side of the bed you fell out on. Like the digestive hitchhikers in our stomachs, most of these organisms cooperate at a mutually beneficial level, in what is termed symbiosis. This web of interconnected life underfoot might be seen as forming the one vast living organism upon which we all depend. A magic carpet.

In industrial monoculture, the topsoil is essentially purged of such naturally occurring life forms, and the whole web is broken, in the interest of controlling and enhancing the living conditions for a single plant. Herbicides and pesticides are introduced, along with synthetic fertilizers. Terminator seeds are planted. All of this sanitizing and grooming of a monocrop might seem eminently logical if it took place on a space station, or in an indoor science lab, but then to the layman it might also seem slightly mad. Because we know better. We already have a beneficial kinship with the land and its complex biology that is antithetical to such practices. Much of our resistance to disease and omnivorous ability to digest a variety of foods comes from many thousands of years lived in proximity to the soil and to domestic animals.

The USDA reports that there were 8,493 certified organic farms in the U.S. in 2005 meeting the uniform standard for organic certification set in 2002. With an average size of 470 acres, which is a little above the national average for that year, these farms work just over 4 million acres of land, which is twice the organic acreage in 2000, but only a tiny fraction of the nation's one billion acres under cultivation. The number of certified organic farms in New York State more than doubled between 2004 and 2007, rising to 735, according to the state Department of Agriculture and Markets. Yet that agency estimates that three to five times as many organic farms in the state chose not to spend the $500 to $1,000 to become certified.

Droves of small farmers in the past thirty years have found and climbed aboard the "organic" bandwagon. A few have experienced wild success, and have graduated to the scale and profit aims of agribusiness. Some have ostensibly fallen off the wagon, over issues and costs of organic certification, though their hearts and practices often remain in the right place.

What is the problem? Organic industrial agriculture may well be a contradiction in terms. At the very least it is an ethical compromise that lies at the bottom of a slippery slope, having mostly to do with scale. When the USDA's organic certification rules were drawn up in 2002, they were largely written by ADM, Monsanto, and other major high-tech players, and it is no surprise that such lobbyist-inspired regulations offered scant accommodation for small diverse farms, and included loopholes allowing limited applications of pesticides and herbicides, and permitting processing and additives long vital to the agribusiness toolkit. In practice the effect of these rules has been questionable. For instance, when manure is used to fertilize fields, it is often trucked great distances from some CAFO (Concentrated Animal Feeding Operation), oblivious of the disconnect this represents. Organic certification often pays cosmetic lipservice to such concepts as "free range" for poultry and livestock, while ignoring the commonsense meaning of the words.

So "organic" may be more a frame of mind than a stamp of approval by any higher authority, which has increasingly been caught napping. If you're buying an heirloom apple at a farmer's market, you can ask face to face how she grows it, what she puts in her sprayer. Chances are you will be able to tell the difference. By the same token, the label and certification on a package from some large corporation in your supermarket might never be able to boast enough to reassure you.

More Than a Glimmer

Of course alternatives continue to spring up all around us. Consider the explosive growth of new farmers' markets across the country, at current count approaching 5,000. And the growth of farmers' marketing co-ops, CSAs, (the acronym for Community Supported Agriculture) which are generating new models of how to share the valuable experience and identity of farming. The Robyn Van En Center for CSA Resources at Wilson College reports that CSAs in the U.S. have grown from 60 in 1990 to 1,150 in 2007. There is also a spontaneous movement afoot at educational institutions, from kindergartens to major universities, to create gardens that will supply those schools with fresh produce along with the experience of growing and cooking it. All these alternatives share in common several powerful impulses: to take back and own the power of feeding oneself, to reclaim an understanding and respect for the sources of food and what growing it entails, and to close the loop between the farmer and the folks he or she supplies with the primary sources of life.

These marketing initiatives are signs of an underlying shift, in the byways and suburbs and hollows, from the vast monoculture processes and products, to a more modest scale of farming, based on the tools and methods of our grandparents, but informed by more recent organic understanding and practices. The values and skills of the past, that employed crop rotation, organic fertilizers, and traditional species of plants and animals, in the end assured diversity and sustainability. It is the ultimate practicality and common sense of farmers that have kept those ways with us.

In farmers' and gardeners' choices of heirloom seeds and plants, and in their thriving interest in traditional livestock breeds, there is connection not just to the qualities and values of the past but to the encoded experience of disaster averted. The DNA of these life forms, after all, bears the success stories of survivors. Such seeds and animals have been adapted over time for marginal conditions, and are prized for their hardiness as well as flavor and productivity. The genetic input of centuries is not to be easily outdone by designer genetics that can create a bruise-proof tomato made to look ripe to the eye but not to the tongue or the touch.

To answer the question of scale, the dismissive accusation that small farmers are dabblers, hobbyists, dilettantes, we might note that there are all kinds of farmers, of all sizes. And that scale is in the eye of the beholder, surely no absolute measure of success. Maybe small only refers to the humility required to do it right, to find and know one's place. In that timeless expression, to hoe your own row. Small farmers share a modesty of expectations that alongside other outcomes allows for intangible rewards.

We might also notice how often in our culture size is connected with delusions of self-importance, self-worth. So that the more successful you are, the more palatial your home, the more imposing your vehicle. The implication seems to be that smaller toys are for kids and hired help. That as a billionaire I would absolutely require an eight thousand pound behemoth to carry me around, both as a sign of status and buffer. We all recognize the obvious safety in the choice of such a vehicle, as well as the implicit cynicism. Never mind right-of-way, in collisions between unequals might does make right.

Signaling the Shift

Since the Industrial Revolution swept across Europe and gained a toehold in the New World, an endless parade of new technologies and techniques have appeared, promising to reward the farmer with greater efficiency and higher yields. More time and money for less effort. But there does seem to be a change, a questioning of the latest high-tech innovations and even a break with that program. What we are calling the New Farming signals a shift in awareness, in consciousness, within three areas: scale, information, and relationship to the market.

First, we can see farmers deliberately thinking small and local. The number of new small operations are supplying a wave of new and burgeoning farmers' markets around the country. Knowing who grows your food has become the antidote to anonymous threats due to carelessness, indifference or worse.

Some of this change is personal, at the level of consciousness-raising. In the field it helps to know yourself, to take stock of your own motivation, your sense of the elusive and often non-material rewards of what you're doing. A farming friend recently told me there are three kinds of farmers nowadays. The first is really a mechanic and heavy-equipment operator, excited to run and maintain the big machines required to work a big place. The second kind of farmer is at home with animals, drawn to the life in order to work with livestock. The third kind of farmer is one who is good with growing plants, working and sustaining the earth. Each of us might well ask, which am I?

Yet all three are necessary on a traditional farm, where these skills and motives should be complimentary, if not interwoven in the daily round. It's only recently that a farmer could prosper working and servicing heavy equipment rather than as a teamster with his team, the one skillset all but replacing the other. As for the second, working with animals fattened and confined their whole lives on feedlots, there is less of that other life to be nurtured and shared anyhow. So it might seem that the farmer moved to work the ground and grow plants would thrive where the other two might all but disappear. In the end we should not kid ourselves that only one kind of farmer is required: all three skillsets are needed, along with many hidden urges for which we may still have no names.

No Telling

Some warm fall evenings you get to hearing things. Rats in the crib at work on the new corn. The hen house unsettled, that something might have found a way into. You get up and poke around, listen to pigeons in the hayloft, barn owls busy at it, mourning doves off in the windbreak. The gathering night seems full of sounds that have nothing to do with your day's work, the harvest. It's confusing. Some ways you feel full, some others empty, some ways you think even now there is no telling if you have too much or nowhere near enough. There are all these others, the lesser ones, who live on what falls between mouthfuls. You wonder what will sell, what will keep, what to even try holding separate from what.

The Myth of Drudgery

There is a persistent myth about farming as an endless round of drudgery on the land, that has been questioned by recent historians. It is true that planting and harvest days can be long, both ends racing the weather, first to get the seed in, then back to the barn just in time. And some farmers, like many other small businessmen, make poor bosses and their own worst employees, when their anxiety about success and failure doesn't allow them any downtime. But there are also farmers who take long vacations and holidays. Who find time to deepen their knowledge, perhaps study geology, botany, organic chemistry, biology, those underpinnings of their craft and livelihood. Farmers whose independence and freedom to schedule their days allows them to enjoy family, other pursuits and diversions, and a full life of the mind.

As Dutch historian B. H. Slicher Van Bath notes, farming in Western Europe, from the Middle Ages to the mid-nineteenth century, involved a workday that began in the field at 5 am and routinely ended by 1 pm. The limiting factor was often the endurance of the oxen and later the horses used for plowing and other heavy tasks, when working in the summer heat. So days started and ended early. The measure of ground that could be plowed in a day, for example, in German-speaking lands, was termed a "morgen"—literally, a morning.

So work in the dirt is not necessarily dirty work, nor is it drudgery, if the farmer is free to shift to some other job. Or take a break when he pleases. Lean on his shovel a moment, revisit the natural world. Which is not to deny there is hard work to be done, but that the days are not as many or as long as we might think. There is nothing intrinsically wrong with physical labor, bending oneself to the task. And there are natural breaks between seasons, windows for diversion and reflection.

Edwin said that his milking operation was worse than a marriage, given those twice-a-day appointments with the stool and pail for thirty-five years, with never a morning or night off. But there are alternatives to stoical acceptance. He could have learned something from recent small-scale dairy farmers who deliberately arrange for a yearly break, take a fall or winter vacation while all the cows are allowed to dry up, then bred to come fresh in a regular cycle.

Living At Large

Thoreau harbored a mild prejudice against farming that seems at least in part to have come from the high cost of land around Concord, how the burden of debt and craving to succeed combined to make farmers hide-bound conservatives, with no room for new ideas. Stooped from that weight on their shoulders. While living at large on Walden Pond, in effect squatting there on Emerson's land, his own tillage included hiring a horse and boy to plow his garden, then hiring them again for the harvest. Though he could have done without both, the leisure for his studies was more valuable to him at the time than any absolute independence.

The Old Thinking

Even with these hopeful signs, we need to honestly admit our failings and weaknesses. With farming there is always danger lurking in the Old Thinking—combative, defensive, self-serving, that says let somebody else take the risk first and show the way. Let someone else quit pumping the water out of the ground first. Let somebody else turn back to dry land farming, and prove that it can be done. Let somebody else blink and take the first step back toward sanity and sustainability. If there's water under my fields, it's mine to dig for. If I let it flow past me, it becomes somebody else's. And my neighbor's gain is my loss.

Hence the Tragedy of the Commons, where any asset seen as held in common, as community property, is overgrazed, consumed to exhaustion, while private property trumps every other consideration. The dark logic being, what's yours is mine, what's ours is mine, and what's mine goes without saying.

It is no accident that the Enclosure Movement spread hand in hand with the Industrial Revolution across Europe, accelerating to its peak between 1750 and 1850. Small cottagers dependent on common lands for pasture and garden sustenance were fenced out by large hereditary landowners, who asserted an absolute economic right, and denied the practices and relationships of centuries. These marginal workers were forced to move to town and work for a wage in what Blake was to aptly label "the dark satanic mills." No wonder so many emigrated.

Blinders

Hallowed by time, some of the failings of the old thinking might be seen as blinders—deliberate limitations of vision that avoid temptation and distraction. But blinders that are useful while working can be obstructions to thinking and planning. We need to lift our gaze and take in the wider view.

Then there is that ingrown variant of the Old Thinking, which is Bottom Line Thinking, which has some disheartening implications, when applied to farming. For instance, why does an animal need a good life, if I stand to gain from its work, from its growth, from its death? In the name of efficiency why not clip the beaks off chickens, pen a hog so it can't turn around, cage cattle knee-deep in their own excrement? What's so special about any other life form, when stood alongside the Glory of all Creation, two-legged and upright?

Blind Spot

Until recently everyone was connected to the soil, which may be why farming has occupied a blind spot at the center of our consciousness, where our enduring need for daily bread intersects with our trust in both spirit and science. So we ignore the obvious questions, such as how we continue to deserve this modest miracle, and what helps it carry us on. Farming likewise stands at the crossroads of our economic, political and social institutions. No matter what the laws say, what it costs, or how it is delivered, we still need to be fed. No wonder that age-old questions of education, right living and social justice radiate outward from its fertile soil. No wonder too that politicians and ministers in ghostly homage still hunker down and talk crops.

Yet religious and ethical leaders of necessity are drawn to borrow the stewardship models of traditional farming and herding as examples of environmental behavior and right living, to counter the exploitative and indifferent practices of corporate enterprise. Farming is a spiritual process and practice, in that an elusive and incalculable non-material gift is involved—life itself. Things we planted grow up and are harvested. We should not be so arrogant or deluded as to take more credit than is due, to our efforts, our timing and vigilance. Our nurturing is miniscule alongside that of the earth.

Telltale Signs

This is the dead time of year. Days are short, nights long, with less
excuse to bundle up and go out, folks and animals mostly penned close
indoors. Though there are still a few chores, and restless walks around
the place, thinking through what's to come, with an eye to the wood,
feed and fodder. These mornings you notice, nothing wants to have to
break the ice to get a drink. You see it even down along the crick.
Having never seen ice in their lives, the young require a lesson. With
the woodpile running low you find yourself stalking the woods,
muttering. Looking for dead standing trees and snags that are no fun
to cut, leaned up against the living, all in disguise once their leaves
dropped. That can break apart, drop a chunk on your head, bite back.
So you look close for telltale signs—the splitting bark, high up
woodpecker holes and frayed limbs, and the fungi low on the trunk.

4. New First Principles

Thinking Out Ahead

The key to all work on the land should be sustainability. Will I be leaving the land, the seed and the breed stronger than I found it, the better to carry on? We have to abandon extractive, exploitative quick-buck thinking, leave that to the drillers and miners and so-called developers. Farmers work to bear life as we know it, its grounding and perennial flourishing, into the unforeseen future.

Still, as farmers we like to kid ourselves. We don't want to think about the consequences of our actions. That farming begins with a disturbance, a disruption of the thriving polyculture that is nature, that we could minimize, dismiss as easy theft, or carry to the logical extreme of rape and plunder. We need to admit our sole intent is to erase what is growing there to plant in its stead something that will support us as well as itself. So in good conscience we need to work to lessen the effects of our actions, heal the trauma and minimize losses to the soil—the erosion, depletion and waste.

Maybe we need to be thinking more like a weed—that is, like any plant perfectly adapted to its environment, which springs up and flourishes without human intervention. At home here and now, a nuisance only because omnipresent. It's a humbling thought that all the productive varieties of plants that sustain us began as weeds somewhere. So at first farming was only noticing what grew, encouraging one weed over another.

But let us return for the moment to the immediate costs of doing business. The recently popular impulse to eat in season what is grown locally—by those who call themselves localvores—is an attempt to raise consciousness about the true costs and hidden assumptions of our agribusiness model. Drawing an arbitrary circumference around oneself, whether of a hundred or five hundred miles, deliberately counters the trend we have been on, to whisk out-of-season crops across the country or around the world. Until recent price spikes at the pump, we have routinely ignored the cost of tillage and transportation powered by fossil fuels, and the further expenses of environmental pollution and global warming.

Measure the Health

Some time early in the Depression, on a hillside of his pasture, Edwin found a spring seeping out of the ground. He formed up and poured a concrete springhouse, then used it to water his livestock. In a part of the world where wells used to be hand-dug, no more than thirty feet deep, now they're routinely drilled to 300 feet and more, chasing a retreating water table. But for eighty years that hillside spring below the house has measured the health of this farm, sign of water still lying close to the surface, that never once has run dry.

The Great Thirst

In the United States it takes roughly 400 gallons of oil per person per year to power farm equipment, which adds up to 17 percent of the nation's current total energy use. Beyond that, oil and natural gas are used to make the fertilizers, pesticides and herbicides favored by agribusiness. Petroleum is also required to process food before it reaches the market. It takes the energy equivalent of half a gallon of gasoline to produce a two-pound bag of breakfast cereal. Which still does not count the energy needed to transport that cereal to market. It is the transport of processed foods and crops that consumes more oil than any other input. And most foods have traveled an average of 1,500 miles to reach the dinner table.

The totals are staggering. The United Nation's Food and Agriculture Organization estimates that feeding an average family of four in the developed world consumes the equivalent of 930 gallons of gasoline a year to produce, process and deliver their food, which is nearly equivalent to the 1,070 gallons that same family burns in their cars.

Let's look at recent trends. The Department of Transportation studied food deliveries to the Chicago Terminal Market, targeting the years between 1981 and 1998. Semi-truck deliveries gained over those years, going from 49.6% to 86.9%, while rail deliveries fell, from 50.4% to 13.1%. Foreign deliveries over the same years rose from 12.5% to 21.5%. And during the same period the Weighted Average Source Distances (WASD) for all produce climbed from 1,245 miles to 1,518 miles.

Can we break it down further? Such calculations are not easy or surefire, though among disputed figures the pattern remains ominous. For instance, the Organic Consumers Association (OCA) estimates that the average person's diet in the developed world—not just the U.S.— consumes 422.7 gallons of fossil fuels each year. Of this total, 31% or 131 gallons is chemical inputs to crops, while 16% or 67.6 gallons is transportation to market. Which leaves 53% or 224 gallons for food processing and fieldwork.

The most neutral way to consider our choices might be as a physics problem, energy in versus energy out. With traditional organic agriculture, depending on animal-powered or human-muscle input, one calorie of energy will grow 4 to 6 calories of food. By contrast, using current agribusiness methods and equipment, it takes 10 calories of fossil fuel energy to produce one calorie of food—which makes traditional farming and gardening 40 to 60 times as energy efficient as petrochemical monocropping. And when we include transportation, we climb aboard the fast track to absurdity. Flying iceberg lettuce from the United States to Great Britain costs 127 calories of fossil fuel energy to transport one calorie of lettuce—and that doesn't include calories lost to petrochemical inputs, tillage, and ground transport on either end. Other British examples include 97 calories of energy to transport one calorie of asparagus from Chile, or 66 calories of energy to fly one calorie of carrots from South Africa.

Most of us are aware of the relative efficiencies of the competing means of transport, though it might be useful to set them out in a row, using ton/miles as the measure—that is, how many tons can be moved a mile on one gallon of fuel. The most efficient fossil-fueled cargo transport is the river barge or ship that can get 500-576 cargo ton/miles per gallon. This compares with 400-436 ton/miles per gallon for rail transport, 100-155 ton/miles per gallon for tractor-trailer rigs, 20 ton/miles per gallon for pickups and light delivery trucks, and 5-7 ton/miles per gallon for air transport. Using a ship or barge as the norm, rail is roughly 80% as efficient, a semi truck is 20% as efficient, a pickup is 4% as efficient, and a cargo plane 1% as efficient.

What are the lessons in these numbers? The first is to slow down, with its corollary, to plan ahead. As speed rises arithmetically, fuel use rises geometrically. The fastest transport is the most costly for the environment, for all of us, while the slowest is the most efficient. We have largely bypassed some of the most efficient means of moving food, in the interest of saving time. Since grapes can't come by ship from Chile, maybe we need to forego such out-of-season luxuries. Perishables need fast transport, or should be grown nearby. Another lesson is that we need to weigh all the costs of our choices, not just those on the price tag. For instance, consider how each of these alternatives requires infrastructure to make its deliveries. Barges depend on an intricate lock and canal and levy system built and maintained by the Army Corps of Engineers, where ships rely on harbors, dockage, and guidance systems. Trains need a rail network, switchyards and right of way. Trucks run on Interstate freeways and on each state's secondary roads. Jet aircraft require airports, GPS equipment, and a grid of air traffic controllers. All of these systems are subsidized by the taxpayer, if not bought by us outright.

Maybe every food label should include a transportation breakdown, detailing miles traveled and cargo carriers employed, with attendant environmental costs. Of course the proposition is absurd—absurd because it is all but incalculable and unenforceable in practice. The status quo possesses massive inertia; big players are reluctant to incur any additional costs, and immediately pass them along to the public anyhow. Still, we are driven to ask how else we might address and mitigate these costs. The gospel of the new Global Economics is that goods should flow freely from wherever they can be produced most efficiently to where they are most needed. In practice all parts of this expression have been manicured into trivial euphemisms: "most efficiently" meaning at lowest cost in wages and benefits, materials and environmental safeguards, "where most needed" meaning to the marketplace capable of handling the greatest volume at the highest mark-up, and "flow freely" meaning moved most quickly at the lowest per-unit fuel cost to distributor and retailer. By any sane measure, globalization is neither natural nor inevitable, much less sustainable. This series of political and economic agreements is intended to favor large-scale production and long-distance distribution over more localized, modestly scaled, culturally and environmentally sensitive enterprises. Which means it should have nothing to do with our food.

Yet we do need to consider our food production and choices in light of how they impact our neighbors near and far. Eating and growing local produce is a humbling recognition of the values tied to one's neighborhood, its soil and growing season. How else can we get to know what is richly commonplace in a region, hence worthy of celebration in a Tomato or Tulip or Artichoke Festival, and what is prized for its rarity. Sustainability reaches far beyond the individual life and its needs, to the survival and prosperity of a culture, a region, a planet.

Speaking of regional celebrations, a local prank comes to mind. At the height of zucchini season in the Northwest, that breed of squash can sulk under giant leaf cover, and before the grower gets out of bed in the morning its offspring are the size of Goodyear blimps. There's a saying, don't turn your back on them. The only recourse is to park one like an orphaned baby on your neighbor's doorstep, ring the doorbell and run.

Join the Runoff

Finally we need to visit the actual bottom line, regarding human life on planet Earth. It's not the dwindling stocks of fossil fuels we've been considering, but the soil itself—whether it remains healthy enough to sustain us. Let's consider the numbers. An inch of topsoil on an acre of land weighs roughly 150 tons. The Soil Conservation Service estimated average U.S. topsoil loss due to farming in 1979 at nine tons per acre per year, or one-sixteenth of an inch of soil per acre. Which in sixteen years means an inch of soil, or six inches in a hundred years. Remember, that's an average. Flat lands may lose as little as three or four tons per acre per year. But there are measured runoffs on some fields of forty and fifty tons, with losses up to 200 a year on steep unprotected slopes. The total U.S. soil loss in 1979 was estimated at over four billion tons.

To balance that loss, how long does it take for an inch of soil to be created or replaced? Under natural conditions—that is, doing nothing but let the land return to the wild—from 300 to 1,000 years. On farms where manure and compost are spread, organic matter is allowed to accumulate, and careful tillage is practiced, one inch of topsoil can be built back in about 30 years. Synthetic fertilizers do nothing to counter soil loss, and as they join the runoff only obscure and complicate the reckoning.

Yet soil conservation has been a hard sell ever since the Depression. Even confronted by the hard facts and long-term implications of soil loss, since the 1930s most farmers have remained unmoved, and only grudgingly played along with the Department of Agriculture, accepting payments for doing the minimum. Maybe we need to throw out the word conservation altogether, scrap the twin sawteeth of saving and spending, since Americans have by and large quit saving for their own futures, much less those of their children and grandchildren. Bumper stickers boldly proclaim, "I'm driving my kids' inheritance." As if one would choose that dead end without a gun to his head.

Yet we need to bank soil for the future. We need to realize we hold it in trust. Beyond the six feet of a gravesite the land is not ours. It will belong to our children in turn, and to their children, on and on. Or to whoever snaps it up after they have squandered their portion. That is, if we possess the foresight to not let it slip through our own fingers and run out to sea, while playing endless games of profit-grab and finger-pointing.

That Elastic Intangible

While speaking of precious resources, we should note our attitude toward time. Our collective will to annihilate time and distance makes them equivalents, our mad rush a denial of the conditions of our existence. Persisting over a lifetime, these demands for instant scene-change might begin to seem the reactions of a surly captive. In any case, we should uncover the secret costs of these impulses—and see what is truly at stake. As a small example, consider the UPS and FedEx distribution model, where all packages going more than roughly a hundred miles by ground network are flown to a national hub, sorted in a matter of a few hours and sent back out by the same planes. Overnight, an individual package may log many thousands of miles to effectively travel a few hundred, which clearly wastes fossil fuel and ignores environmental impacts, in the interest of saving that elastic intangible—time.

Our conscious perception of time is all but absolute. As if it were actually ratcheting gears, or hourglass sand, always running. Yet we can't stand to watch the clock, and are only intermittently aware of time's passage, only notice how it slows or stops under great duress, sudden relaxation, or the perfect diversion. But then the instant the music ends we snap back to the clock. We confuse moments with time, and behave as if all had an equivalent meaning and value. Yet what is the $1000-per-hour surgeon worth while cooking his dinner, or brushing his teeth? Or the billionaire banker gardening?

The small farmer knows the difference between a moment and a minute, and is unlikely to confuse the two, or squander either. Inhaling the scent of his efforts, measuring his days against the tasks at hand, he does not inhabit a "time is money" continuum. There is sustenance in the cycle of the seasons, in their returns a renewal. His calves take their time being born, his crops make when they're good and ready.

Feet on the Ground

We may be cursed with the mindset that assumes every big problem requires a big solution. Some years ago in the Northwest the sport of tug-of-war enjoyed a brief vogue. Under local rules each team was given a weight limit. All the competitors used a standard eight-man team except one—the Port Townsend Centipedes. Reasoning that the secret to success might be traction, they recruited smaller athletes, and fielded a ten-man team that made the same weight limit, but put more feet on the ground. And for years walked away with the prize.

Which may be exactly what will ensure that our food is safe and nutritious—more feet on the ground. The more farmers, the wider the net of our connections, the more varied and intimate our supply lines, the less likelihood that any one act or event could disrupt us. And it is not just farmers who are networking to improve food safety and freshness. The best chefs and restauranteurs have long built their culinary successes on personal connections, knowing the suppliers and sources of all the produce, wines, meat and fish they prepare. Even visiting fields and pens and fishboats to see for themselves. They are also used to paying a premium for that access, that personal touch. How else can one ensure freshness and flavor, and maintain a competitive edge? This thinking is being adopted by many families who also crave the best and safest food. They step up and get involved, learn all they can, and keep their eye on the prize.

Off the Old Grid, Onto the New

With the New Small Farmer, then, what has changed? The first clear shift might seem to be away from isolation, the limitations of living and thinking remotely, at the end of the road. With electronic and print media widely available, there is potential access to the good minds and focused energies of others working to solve common problems, readily available at a minute's notice. In seconds we can do a web search to locate growers by name, region, and crop. We can survey CSA farms, organic growers and U-Pick operations, sort by location and specialty.

Of course these are just starting points. There are problems inherent in the Internet, which stands open to all but harbours a potential for endless distraction, deception and drivel. The new connectedness is exactly what we make of it, and depends on each of us maintaining intact what Hemingway liked to call our built-in crap detector. Along with the best of the past, new ideas and techniques could be shared quickly, talked over, questioned and improved. That is, given the requisite attention, selectivity and self-control on the part of each of us.

The second change is a new union of the small farmer with those who eat what is grown, a common purpose with those who share the bounty that fairly cries out for a new name beyond "consumer." Many such unions aim to become thriving partnerships, and can be. The small farmer involved in the CSA (Community Supported Agriculture) movement today develops a working relationship with his subscribers and customers, based upon common interests. Both care enough about the worth of food that it might be demeaning to call the bushels of sweet corn or crates of melons a product bought and sold.

Both understand at some level the right living and meaningful effort implicit in the crop, the compliment implicit in the exchange. I feed you what I feed myself.

The advantages of the CSA model are many. While building community it avoids the risk of the small farmer growing produce that may or may not be sold before it spoils. The small farmer's produce would otherwise have to be marketed alongside the usual seductive displays of agribusiness. In the CSA partnership there is little or no waste, a natural connection with the seasons, and every subscriber is given opportunity and incentive to try unfamiliar fruits and vegetables. Opportunities to volunteer abound, with subscribers learning informally and getting to know their neighbors. And most importantly, the financial burden and risk are shared: the farmer knows how much he needs to grow, can vary his offerings if one crop should do poorly, and receives payment in advance, so that at the start of each season he is not forced into debt.

There is also an ongoing revolution in the relationships between small growers and larger retail outlets. Direct Store Delivery (DSD) by small suppliers accounts for a growing share of produce in local markets, which have increasingly welcomed small farmers. In 2006, the organic produce chain Whole Foods Market announced that it was directing its regional buyers to focus on obtaining local products, and expanded their suppliers to include 2,400 independent farmers for their 275 stores. That year small farmers accounted for 78% of the fresh produce sold by the chain. Though Whole Foods is definitely Big Organic, their announcement shows that they've at least read the handwriting on the wall, and are attempting to adjust.

In a new world struggling to be born, we are all by turns feeders and fed. Fellow farmers. If we allow only the cash exchange and marketplace tussle of capitalism to prevail, we may have to swallow the heartless truth that there are careless or indifferent poisoners in our midst, who are largely untouchable, who stand to profit by our sickening and can never be held to account.

The New Calculus

Luckily there are new ways of figuring one's place in the scheme of things, and computing true costs. Consider the contrasting economics: with a team you might grow and feed your own oats and hay, instead of purchasing fuel to run your equipment. Growing the feed for your team offers clarity of purpose, little cash outlay, and minimal transportation costs. But it does take additional time, effort, and forethought. Instead of the turn of a switch, there may be up to an hour on either end, feeding and harnessing, walking to and from the field, hitching up and cooling down. But the logic is inescapable: the working team helps feed itself. By contrast the tractor-powered solution is quick and dirty, convenient and thoughtless. It contains no line items for airborne hydrocarbon emissions, noise pollution, or global warming, though we now know full well they are part of the account that's come due.

We could go on to contrast the economics of breeding one's own draft animals, which for many is a distinct and pleasant possibility, with the building of one's own tractors from the raw materials up, which for most is a patent impossibility. One might as well speak of drilling and refining one's own oil, to fuel one's own tiny asteroid.

Measuring Success

You might happen to be raising a bunch of calves right when the price of feed doubles. And you might have grown enough feed to make you immune to this shock, or you might be caught short and have to sell your animals to avoid further losses. Or you may have finally found the perfect animals to build a herd, and feel you need to hold onto them no matter what. You may have options or not. Either way, the more you care about your livestock, and the more you take the long view, the more complicated the calculation, and the less it can be discounted, oversimplified, dismissed as profit and loss.

So there are good years and bad, not just ones where the numbers work out and when they don't. And for the farmer there are even longer and less obvious measures of success. Fertile fields well treated accrue with the years. The success of the soil or the breed is an ongoing benefit to be savored as it helps lift and carry one along in the work, not a momentary triumph but a quiet undulating joy.

Remote Control

Speaking of markets great and small, we need to reconsider the logic and ethics of trading in commodities futures. And perhaps instigate the writing of new laws. What kind of a market is it where an investor who took none of the risk to raise a crop can lock in the price and corner the supply of soybeans or corn or pork bellies for years to come? Of course any successful producer wants to know what his future costs will be, in order to limit them. But next to the farmer's risk of total loss, the commodity investor's risk seems trivial, that the day-to-day price might not continue to rise, in the face of an absolute certainty, that an ever-increasing population will generate more mouths to feed. This past year on Wall Street several large investors managed to buy up all the surplus grain storage capacity in the United States. Setting up a turnstile so they could take a toll of any abundance that must perforce flow through them.

We all have abilities and offerings to share in a marketplace that is trusting, informal and sociable. Every advantage need not come at the cost of some other's disadvantage. At the extreme we have seen the consequences of a voracious and unregulated market, where it is open season on trust, where government regulations and their enforcement begin to seem only laughable games for the cynic who in the end will be forced to grow and provide only for himself.

Which brings us to a small but necessary corollary. That farming on a modest scale can provide a rare antidote to cynicism, that pushing seeds into the ground is at its heart an intrinsic act of hope. Despair can have no part in such a life-giving, life-affirming effort.

One Seed

Beeline

The bees in the orchard are working cherry blossoms, an endless melodious thrum in the air through the sunny late afternoon. Overburdened, their anklets golden pollen, the bees labor by on a drooping trajectory. Where I stand it is fragrant, alive, with my back to the hive, facing the biggest old cherry—proud, like a bride in full bloom. Every now and then something bounces off my chest, tumbles to the grass, fusses and struggles to make it back aloft. Finally I notice what's going on, that I am just on their beeline, that they are adrift in the faintest cross-breeze. I take a couple steps to one side, and watch them go lumbering past.

Spreading the Risk

It is a truism that farming is among the most dangerous occupations—perhaps most dangerous if it is taken into account that most other dangerous occupations include training to cover their hazards. But apart from the Ag colleges that until recently confined themselves to teaching the science, and otherwise mostly touted the dubious accomplishments and goals of agribusiness, farming has no institution to inculcate its skills, to create and maintain its standards, as do firefighting and law enforcement. On the farm threats to life and limb are somehow factored in with all the other risks, of livestock disease and crop failure, of flood and drought, fire and pestilence, whether due to the unforeseen consequences of one's own choices, or to acts of God.

Farmers deserve and would welcome insurance of all kinds. Which agribusinesses can afford, or in times of trouble rely on government programs, loans and bailouts. But at the small end of the scale, precious little is currently offered. The small farmer should not be compelled to become an employee of some larger concern, forced to divide his successes and blur his identity in order to garner protection.

Perhaps we need to return to the philosophical and ethical origins of insurance as a shared risk, where the community accepts part of the burden on each individual. Why? Because lightning could as easily have struck my neighbor and didn't. Because my cows sickened and recovered, where hers died. It is an admirable notion, at the heart of what we mean by community. Yet as soon as the notion of profit arises, and insurance is issued on a business model, there appear acceptable and unacceptable risks, with some accountant empowered to draw distinctions that will deliberately exclude some part of the potential pool of those insured. And blame is apportioned and affixed. None of which would do justice or provide comfort to the small farmer who stands in need of the same even-handed, healthy and watchful community that his produce and livestock support.

Wearing Many Hats

Most small farmers of necessity wear many hats. The need to secure health insurance for the family or a bank loan will often require that there be a wage earner holding down a traditional full-time job. And with seasonal work, and multiple sources of income that tug in different directions, the farming family endures a range of opportunities and deficits, with a drive to farm that occasionally must clamor for attention while it competes with other needs.

One of Edwin's farming sons worked a full-time job for forty-five years, as an environmental water specialist for a large soap company, a job for which his farm upbringing eminently fitted him. He farmed his own place and that of his father at nights and on weekends. Then at 65 he returned to full-time farming in the daylight, what he calls the one job you don't retire from.

So we might think for a minute of the small farmer's life on the margins, to which he or she is often relegated by the tax code, economics, and family obligations. Perhaps we can learn from those other marginal dwellers on the land, studying the kinds of activities that embrace or exclude them.

Life on the Margins

It is easy to be a lover of nature in the abstract. To pursue a schizophrenic existence where you work in town and on weekends take a hike in the parks or beaches or mountains. Cherishing some imaginary "wilderness" that doesn't exist except as the kind of nature museum many parks strive to offer their consumers, reflecting the values of an urban world rigidly based on mortgages and platted deeds, where undomesticated animals are not tolerated, viewed as threats to health and property. Where we are cautioned to keep to the trails and not pick the flowers, though at home we might indulge the impulse to fill a bird feeder and drown a potted plant.

Agribusiness likewise has no place for what are regarded as trespassers, interlopers, thieves and pests. Yet, perversely, the raising of vast monocrops creates conditions for infestations of insects, the predations of birds and deer and other unwelcome visitors. By contrast the small farmer, with only a few acres of any one crop, incurs little risk of nurturing a scourge or pestilence of biblical proportions. He meets nature not as a possession, nor as recreation, but as a daily negotiation, as but one among the many lives involved. On the small farm, along fencerows, in gullies, around woodlots and pastures and on steep hillsides, wildlife is accepted, tolerated, if not welcomed.

Why? Because for the small farmer, signs of nesting and thriving creatures, both resident and passing through on their migrations, are reminders that these borderlands and edges represent continuity; that there is a necessary balance and interchange in the treetops and watersheds that eludes rigidly drawn property lines and defies conformity; that the land itself asks enough ground cover to counter erosion, enough scrub and undergrowth, whether unsightly or not, to anchor the here and now. The willingness of the farmer to share and not begrudge a little to his smaller, less visible neighbors betokens a healthy attitude toward the land, a genuine stewardship. It is not all mine, nor can it ever be.

So the farmer engaged in a skirmish or waltz with those other life forms may make the best environmentalist. Every day he sees the competitors for what he grows, and is in touch with the cooperators in his task. He has to act with restraint and clarity. He has to learn how to share and play well in order to get along, to sustain his small part of the whole.

From the Leavings

All day in the field mowing hay, we watch black vultures spiraling over the far end of the pasture. If we had any animals there, we'd have gone to look first thing. But there is a spell to the shuttling mower in step with the harness jingle. As it is, late in the day we find a yearling doe that something has brought down. It is hard to tell from the leavings just what happened, and by now there are so many different tracks—fox and coyote, raccoon, mink and whatnot—we let it be where it lies.

Countering the Sprawl

Driving the back roads, we can often tell where the small farmer is alive and well, where there is no screen of new houses backed up against tilled fields and pasture. That openness can be a signal that here the frontage hasn't been sold off for quick and easy money. We can look over the fields to see if there are breaks of trees and shrubs, signs that the farming is not done "fencerow to fencerow," excluding toeholds for wildlife. Clues that a certain openness extends to what is being grown and how the work is done. An endless wall of new upscale homes along a country road in good farm country may signal a run-down neighborhood—for farming, anyhow—one where the farm work might be barely tolerated or even impeded by newcomers who will need to learn to be neighborly.

This sprawl represents one of the false choices offered to Americans since WWII—letting open spaces solve our social problems. The flight to the suburbs has been a movement to seize upon and squander but one part of the rural life—its elbow room—draining both the rural landscape and the cities of their vitality and sense of community. Of course the economic engine made it all possible, featuring high-paying jobs, a dependence on the freeway system and cheap oil. Increasingly people live at one freeway exit, work at a second exit, and shop at a third. In its apotheosis, the family castle—appropriately dubbed a McMansion—sits at the center of a spacious green plot, with its neighbors held equally distant. All the rest of farming and ranching, connected to its purposes, has been manicured out of existence. So that what we are left with is a sanitized lifestyle of lawn mowing, bug-spray, pets, and long driveway walks to the mailbox.

How You Gonna Keep 'Em Down on the Farm

One condition of small farming we won't be going back to is its reinforced and often corrosive isolation. Contemporary farmers know the wider world, and though they may choose to hold it at arm's length, there is ongoing access to new ideas and alternatives. The Internet, cell phone, television, radio and other media have made vast inroads into rural remoteness and reticence, though at the same time we have to note how much of that abundant information can prove manipulative, deceptive, or worse. More than ever the messenger—in the case of the Internet always invisible—must be viewed with a skeptical eye. And though the situation is changing, the outreaches of the Department of Agriculture, as well as the forays of the major agricultural colleges and universities, continue to be skewed toward monoculture and its limited view of rewards. Such values as self-sufficiency and sustainability are often still ignored or met with scorn, and are mostly not practiced or taught.

Still, there are signs of a shift. For example, The Aldo Leopold Center for Sustainable Agriculture at Iowa State did a 2008 study exploring whether it is more energy-efficient and environmentally viable for farmers in CSAs to deliver produce, or for members to make pick ups themselves. (Delivery won by a considerable margin.) We could use such thinking to help refine options and guide farming choices. Agricultural schools could take on a whole new role, to awaken and focus environmental concerns, to study and honestly report their results.

So we are talking to one another, speaking our minds and learning where to turn for answers. We need to hear fresh solutions to problems that challenge our own experience. We need to be jostled, stimulated, not lulled to sleep in the face of difficulties.

Doing More With Less

The new farming is based upon improvements in training, techniques, and tools. Some of these are radical departures, but most are adjustments and refinements all but invisible to the casual passerby. The inventiveness summoned in using and recycling old technology can reach toward art, in its economy and awareness of limitations— forever doing more with less. So traditional farms that look from outside like "period pieces" or "living museums," on closer inspection might prove to be anything but. The immediate and eternal concerns merge when trying to find the best way. Let's briefly consider in turn the new training, practices and equipment of small-scale farming.

Spare the Rod

People who work the land with animals today do so because they want to. A lasting side-effect of that shift from necessity to choice has come at the heart of the practice. In the training of horses and other draft animals, gone is the old coercion, the "breaking" that used to be a precondition for controlling all horses. There have always been humans who could get the best out of work animals, who in place of intimidation instinctively made use of a smarter and more humane approach. But recent methods represent a shift analogous to that in child-raising, where up until fifty years ago most parents believed in "spare the rod and spoil the child," and spanked their kids, supposedly for their own good. Based on a better understanding of the animal's senses, feelings, behavior and intelligence, training is not just one-way, but involves the teamster in a working partnership with the animal, in a practice that for both values clarity of intent, understanding and patience. And this humble and observant approach has often meant a lifetime of shared effort that is its own reward.

With a Team

First and last there are rhythms to speak of, slow steps to the pull, measure in the muscle, feel of scale. There is the forethought of hitching up, anticipating the day, meeting head-on what needs doing, sharing a piece of it as work, then the winding down after, the cooling and unharnessing, in release the impatience to be fed. There is quiet, mostly non-verbal contact. There are blinders and horizons, intangibles warm to the touch. There are values that remind the human of his or her place in the scheme of things. That by the roadside often will come a stray mouthful. That resting the team hints at resting oneself.

With a team you are less likely to overwork if you are subjected to the same conditions, the same heat or cold, the same biting flies or downpour. There is another logic that takes over. When you come to a hill with a load, you know with the team every little bit helps, so climb down and walk alongside. You may even wade the wild stream first to test the footing, as well you should.

Though as teamster you take the lead, you grow to savor the partnership. Here are more sets of senses in play. And often more safety and sense to the group, its full attention gathered for the word. In the quiet clear air together, in time to do something about it, you may be able to see and hear—even smell—trouble approach up ahead.

Anything But Routine

Some of the latest techniques might be classified as new-old, since much has been tried and forgotten, and only recently rediscovered.

Such as intercropping, where more than one crop is grown on the same soil at the same time, with enhanced results. In the tropics beans and peppers have long been routinely planted in the shade of taller bananas and date palms, in a technique that is immemorial among Mesoamerican peoples. And imitating the Iroquois and Cherokee practice of centuries before, Appalachian farmers would plant pole beans after their corn was up, and let the bean runners climb the stalks of the taller plant.

These new-old ideas can come in disarmingly simple packages. For instance, longtime author, horse farmer and editor of the <u>Small Farmer's Journal</u> Lynn Miller has for years been suggesting that instead of combining their oats, farmers try mowing them a little green and stacking the sheaves head-in. Then feeding them stalk and all to the horses. Which cuts down on waste and work in the field, and seems to suit the animals just fine.

Some of this new thinking rejects easy givens. For example, in parts of the country with the right soil conditions, horses are worked unshod, though their hooves are kept carefully trimmed. And some farmers have abandoned barbed wire altogether, in favor of smooth heavy gauge steel fencing, rejecting that chance of injury to livestock.

We might admire the elegance of lightweight greenhouses built on long narrow sled runners, so they can be towed out of the way once plants are well up and the season has advanced. Then with another crop planted in late summer they can be towed back over the beds in fall, to extend the growing season. Naturally well anchored to the ground when they're not being moved.

Many orchards profit by new pruning methods: not just a single heavy midwinter pruning once a year, but two prunings, both lighter, adding a "fall" in midsummer after the fruit is set, to reduce water sprouts and enhance fruit without stimulating plant growth.

One of the largest shifts in thinking has yet to be mentioned, reversing the feedlot treatment of livestock. In the new calculus, with appropriate fencing and pasture included as part of crop rotation, on every field in turn each animal grazing feeds itself, and serves as its own manure spreader, saving the farmer the energy and effort of both tasks. As an added benefit, streams are no longer polluted by the concentrated runoff from feedlots.

Carried a step further, under the heading of "grass farming," there is a good deal of new thinking regarding the flexible use of pasture, with portable electric fencing that lets livestock be moved every day, spreading their manure while stimulating the growth of grasses and preventing overgrazing. And in the same fields, bottomless chicken coops on wheels, whimsically dubbed "chicken tractors," allow birds to graze and forage for bugs as well. Moved along behind the cattle on a four-day rotation, the chickens get to eat fly larvae from cow manure, for them a delicacy.

There is much new exploration of no-till and low-till agriculture, for instance planting grains in the root mass of a hay field after harvest, to reduce soil disruption and runoff. Since no-till can still be counterproductive to sustainability if the farmer relies on herbicides to control weeds, this work goes hand in hand with experiments in mulching, ground cover and crop rotation, that diminish the need for weed cultivation, and retain moisture.

In land use there are creative new models attempting to rein in the suburban impulse, substituting an agreement to hold and work a farming commons surrounded by individual homes. This legal construct seeks to recreate what old farmers might recognize as a village, or what city folk recall as an ideal neighborhood.

Foremost under the heading of long-term ideas, the search is on, led by Wes Jackson and his team at The Land Institute in Kansas, and joined by the Rodale Research Center in Pennsylvania and others, to develop grain and seed crops that don't need to be planted every year—that will reduce tillage and nutrient runoff, and offer farming a new paradigm for sustainable agriculture. Using perennials saves soil, energy and field labor, and increases biodiversity. Experiments are ongoing with such promising crops as Eastern Gamma Grass, sorghum, Giant Wild Rye, Wild Senna, Illinois Bundle Flower, Maximilian Sunflower, and Curly

Dock. And several strains of the Incan perennials Quinoa and Amaranth are already being grown commercially in North and South America.

From Italy there comes the idea of Agriturismo, a government program deliberately created to support small farms that offer Bed and Breakfast accommodations for travelers, and authentic rural vacations. The law's provisions reward the serving of locally grown foods, and offer help restoring old buildings and maintaining traditional small-scale farming methods.

One other note, under the heading of new-old thinking, is the history of a neighbor's brush with catastrophe. Cuba's story of its "Special Period in Peacetime" since 1991 holds many lessons for us, large and small. Following the collapse of the Soviet Union, the transformation of Cuba's farming offers a model of resilience, of swift and sure-footed change, of response to threats and vulnerabilities we share. When 80% of both imports and exports evaporated nearly overnight, and oil imports dropped by 90% due to the abrupt end of the oil-for-sugar trade with the defunct USSR, the Cuban economy was abruptly forced to reinvent—and feed—itself. Without fuel or petrochemical inputs for the modern large-scale agriculture their economy had grown dependant upon, the country's about-face was painful, at moments verging on starvation. The average Cuban lost twenty pounds that first year, and schoolchildren were routinely malnurished. But with a redistribution of arable land and a rediscovery of the skills of subsistence farming, the Cuban people survived. Over the past twenty years a substantial work force has taken up small-scale sustainable agriculture, tilling plots manageable by animal power, mostly oxen and cattle. And the whole culture has pitched in to help feed itself, relying on vacant lots and rooftop gardens, and improvised local markets.

Wherever it takes root, at the heart of the new farming lies an awakened respect for local knowledge, how it is gathered and kept. How a budding community of farmers note and share when a particular pest arrives or hatches out, what the optimum planting dates are for different crops under local growing conditions. What changes year to year. What grows best here, weed and seed alike.

Right Tool for the Job

Finally, turning to hardware, consider another invisible change that appears practically everywhere. I have hardly seen a horse-drawn wagon or carriage built or rebuilt in the past 40 years that has not been fitted with state-of-the-art wheel bearings and hydraulic or electric brakes. Even historically accurate reconstructions have been silently retrofitted, to decrease rolling resistance and give teamsters more control.

In a parallel development, consider the forecart. Resembling a chariot, this modest two-wheeled steel platform has revolutionized animal-powered farming, with its safety and simplicity. The teamster gains the advantages of a stable, comfortable seat and the increased control of hydraulic brakes, that offers a hitch behind for any traditional tool in the field. There are even forecarts with power take-offs, both ground-driven and powered by a gas engine mounted on the platform behind or alongside the teamster.

The list of new equipment includes horse-drawn no-till drills that work every bit as good as the massive motorized ones. And horse-drawn ground-driven hay balers, and horse-drawn in-line mowing machines designed to cut a narrow swath, for mowing vineyards, orchards and gardens. One bright new idea on the scene is a treadle-operated, foot-powered grain thresher designed and built in China, that, along with a grain-mill, would allow the farmer with a small patch of wheat or other grain to supply his own needs for the table. You hold a handful of heads in the top while you pump the treadle; the arms knock the grains loose, and sort them from the chaff.

And there is more. From the latest drawing boards, out ahead we look forward to solar-powered electric tractors, which promise clean and quiet new sources of motive power that could revolutionize field work. Amid these solutions to old problems, we see no blind reliance on tradition for its own sake. There is a marvelous inventiveness springing up to meet perennial farming needs. Whatever you might have dreamt of, there is somebody out there working on it. Or somebody waiting to try your new idea out in the field.

to Another   79

Jockey Hook

The trim, middle-aged lady was leading out the tallest horse I had ever seen saddled, near 19 hands, and since the stirrup was touching her ear, I tagged along to see how she was going to get aboard. Maybe spot the stepladder, give her a hand up. But then someone I knew approached and we started talking, and I never did get to see how she did it. Though I did see her in the ring atop that magnificent palomino gelding, an outsized cross of some kind, maybe part Clydesdale or Shire. And how smoothly she put him through his paces, how comfortable they looked together. Clearly they had measured some miles.

Later I watched that same woman loosen and toss down her saddle, which was one of those new plastic ones, feather-light. Using a traditional saddle fit for that horse, she might as well have tried tossing a green bale of hay to the rooftop. When I mentioned her to a teamster friend, he smiled and described an old tool used to extend one stirrup so a short person could mount a tall horse—the jockey hook.

The Myth of Obsolescence

There was that transition after WWII, when horse-drawn implements had their tongues cut, split and pierced for the tractor drawpin. Tractors weren't so big then that plows, mowers, rakes, wagons and all manner of old gear couldn't get a new lease on life, eke out a few more decades. Of course the old-timers could lubricate, sharpen and adjust, and knew you shouldn't mow or spread manure faster than the horses would have worked, or you'd burn the old thing up. Now, thanks to the reprinting of repair manuals and revival of maintenance skills, many of those tools have lingered long enough to be rediscovered, reclaimed from retirement as yard art, and put back to work in the field.

When we call some tool or method obsolete, we simply mean it has fallen into disuse, which is not to say it has become useless or has ceased to work. Something new has come along, all shiny and seductive. Usually the old way continues in the hands of some, alongside the newer solution, in what amounts to an informal test over time. Humans are known for this, harboring and even encouraging such redundancies. After all, it is how our redundant tripartite brains are wired. We continue to use an array of different technologies alongside one another, ways of working that in some cases span thousands of years.

Often the new technology, the new tool or method is so vastly superior there is no looking back. But that is rarely the case. The wood chisel holds a place in the kit alongside the router. Then some older technologies are immediately or eventually missed; they are found to have contained such economy and elegance that they find a niche, and retain some adherents forever. Look through any drawer full of can openers or writing implements. Or through any carpenter's chest. Even deep into the computer age, some of us find ourselves attached to the fountain pen, and some to the lowly lead pencil. Few of us might choose an eighteenth century tinderbox to light our fireplace, but it remains a dependable solution, and holds a place in the spectrum of inventions that includes flint and steel, the safety match, the Zippo lighter, and the disposable Bic. And despite sharing the home with a gas furnace or electric baseboard heater, that fireplace for many remains the center of the home, the irreplaceable hearth.

Further, the greatest refinements in some technologies have come only after they were regarded as obsolete. Consider the compound bow, for example; its system of lines and pulleys that allow one to hold the bow fully drawn at much less than its maximum pull, for increased accuracy in the release. Or consider advances in sailing. Due to improvements in design and materials, boats sail much closer to the wind now, and with greater speed and safety than square-riggers ever had while routinely transporting freight and passengers. And some of the refinements of sailing should be considered new ways of thinking— such as self-bailing cockpits, that will keep a boat from being filled and sunk by a mishap.

Delusions about scale, along with supposed obsolescence, can dog even elegant solutions. Windmills to generate electricity were considered old technology by the 1950s, too small and intermittent a solution to meet the country's large-scale power needs. Dismissed alongside solar panels. But somewhere along the way toward $5-a-gallon fuel the original objection was turned on its ear. Rather than thinking any one windmill or solar panel too sporadic, a power source too easily interrupted, folks began to envision a large and widely dispersed network of windmills and solar panels to meet our power needs, which after all are also widely dispersed and intermittent. We should be able to find a way to make inputs and outputs average out. Somewhere the wind is always blowing, the sun shining. It just needs to be gathered, stored, passed around.

Of course farm work with draft animals is the quintessential outmoded technology, supposedly bypassed by more recent solutions. Dismissed as an exercise in nostalgia. Yet animal powered farming retains its adherents due to a quiet elegance, coupled with an innate practicality. How else can one grow his own fuel, and keep on hand an endless supply of fertilizer? How else always work in company?

Need Once Met

Across the gravel road at the end of our neighbor's field sat a moldering Model T Ford tractor. It was not one of those kits that appeared in the 20's and 30's, but a crude homemade rig. There was no body behind the firewall, no windshield, just a crate for the driver to sit on, and a larger box full of crick rock bolted over the rear axle for the weight. The rear wheels themselves had been doubled, using long bolts and some kind of spacer blocks, and for the traction a large diameter rope had been spiraled though the wooden spokes and around the pairs of tires. To compound the lower gears, a Model A transmission had been welded in behind the other one, and a sturdy hitch spanned the frame rails. These weren't folks to bring equipment in out of the weather, or hide their failures. So it probably sat where it had quit, a monument to a need once met by a flurry of ingenuity, that now rested undisturbed.

Weed-Eaters

We have recently seen farming know-how borrowed and turned to other uses, blurring easy distinctions, showing how versatile and effective the old ways can be. For instance, in many municipalities herders with goats are being hired to "mow" stretches of steep roadside and other awkward locations, confining the animals with orange plastic sheep fence that can be set up and moved in minutes. These goats can make quick work of a blackberry thicket, and offer a quiet, safe, organically friendly alternative to the usual weed-eating, with its environmental waste and pollution.

Or imagine the blind woman who has adopted a miniature horse to lead her, and is advocating their use as guide animals. Docile, strong and sure-footed, as trainable as any dog, and as easily housebroken, she claims her horse will outlive three dogs, at a considerable savings in training.

An outsider visiting Amish country might wonder why most Amish buggies appear to be pulled by the same identical lean brown horse. When I finally thought to ask, I found out that these are recycled trotters and pacers from the harness racing circuit. When they get too old to be competitive and would otherwise be sold off as dog food or pets, these quick and well trained animals are snapped up by the Amish at auction, to live long, useful lives as people-movers.

When I was a boy, the milkman in our neighborhood had a horse-drawn wagon. The horse knew the route as well as the driver, and would follow him down each street, stopping at curbside, watching closely as the man made deliveries and picked up empties, moving on without so much as a word. I can imagine in some not-too-distant future how my mail carrier on her rounds might be shadowed by a burro or llama. When the scarcity of oil has forced not just mail carriers but all of us to spend more of our days on foot, and erased the profit margins of UPS and FedEx, such delivery animals and carts might again find their place in a world peopled by pedestrians and bicycles, all but emptied of fossil-fueled cars.

Life in Harness

Looking over the racks of harness offerings at auction a couple years back, I asked one old teamster how good these new nylon and polypropylene and biothane harnesses were. He said he didn't know. I must have looked puzzled, wondering why he was standing there fingering the same stuff I was, but then he went on, said he couldn't tell yet if any of them could beat leather. Yet. Said he bought his first leather harness in 1942 and had only replaced one tug and a couple of quarter-straps since. Said he didn't fuss with it much, only wiped it down and oiled it maybe once a year. And kept an eye on it. "When it finally wears out, I'll let you know what I get," he said with a chuckle, and hobbled off.

5. Steps To Take

Pedal to the Metal

What if you're caught in the very dilemmas we've been talking about.
Whipsawed back and forth. It can be tough to know where and how
to begin to change course. But if you find yourself charging at stop
signs, at least take your foot off the gas. In other words, don't keep
adding your momentum to the wrong direction. Begin where you can,
and do what you can, with what seems most under your control.
Working days and farming nights and weekends for some can be a
special kind of hell. Dependence on a cash crop that you dislike may
have you stumped for a while. Or you may feel stuck in a predictable
and uninspired round of the day-to-day. If so, remember, nothing is
forever. Let your frustrations offer their counsel but not overwhelm

you. Plan and experiment, get ready to grow something else that you are more interested in, and accept the shift in income and learning curve as necessary expenses. Dependence on a huge bank loan or mortgage may feel like dependence on foreign oil—insurmountable in the short run. But farming is about the long term linked to the day-to-day.

Sausage-Making

What is the sausage-making part of farming, that no one craves to see? Is it the hopeless vigil over animals you've cared for, grown to know too well? Or is it the actual slaughter, rendering all but the squeal? Is it the sleepless night's worry over vagaries of weather, or is it work like brush cutting and fence-mending, that holds no immediate payoff? Or is it the horns of that age-old dilemma, the cost of land and equipment versus what it might let you earn in a lifetime? Each of us carries around a little personal hell we could use a hand with, that we hardly think to say a word about.

My Other Truck

You probably know most of what to do already. Join, band together. Seek tangible ways to be neighborly. Slap on a thousand bumper stickers that say "Dig Your Inner Farmer." Or slap one on your own bumper that says "My other truck is a horse." Don't fear that you'll lose your precious identity as a stoical outsider by helping others and being helped in turn. Find a mentor—or be a mentor. Ideally you could do both. Teach and learn all you can, formally or informally. Locate farming or gardening companions you can visit—go see what he's planting, what she's got started in her cold frame. Share information, loan and borrow the best that surrounds you. Learn to sharpen your own tools, keep them oiled and bright for the next use. Jostle and challenge your most cherished notions about what works. Start a greenhouse to stretch the growing season on both ends. Try mulching your crops to hold moisture. Once a month take a fork to your compost.

One Seed

Develop a long reach, testing the latest ideas and innovations out there, alongside the best from the past, seeking the sturdy and simple, rejecting the finicky, fragile and dubious. Widen your sense of who might make a good farmer and who is a potential farming ally—after all, nearly everyone can garden, and everyone has to eat. Try invading someone else's turf. Many small farmers are striking deals with restauranteurs, challenging each other by swapping wish lists of what to grow and what to cook, and are recruiting help from suburban teens and retirees to pick crops. Visit your local blacksmith or machinist with a sketch of that new idea. Take the long view, and work toward what waits beyond the horizon. Vote your grandchildren's future. Save yourself and those you love, in widening circles outward.

Stuck Around

About twenty years ago the old folks put up their electric ice cream maker, and went back to the hand-crank model never worn out or given away, that somehow stuck around. It seemed like here was a lesson being lost on the grandkids. What their schools like to call a teachable moment, out on the porch after dinner, as the evening breeze would come on, stealing the heat of the day. That what with the elbow grease and sore muscles, the waiting while the bucket's passed around, all that playful banter dumped in with the fresh picked berries, cranked and cranked till it can't get any harder and won't keep another minute—somehow real ice cream shouldn't happen any other way.

Better Angels

Its first spring in the White House, the new First Family broke ground, planted a garden of 1100 square feet, including 70 varieties of vegetables and fruits. It may be dismissed as a symbolic gesture, but there is actual food being grown, tended and eaten at the seat of power. So that patch can't help but serve as a model of diversity and scale, a way of keeping in touch with what matters.

We have been making the implicit assumption all along, that small-scale farming represents an appropriate response to the environmental and human troubles that are bearing down upon us. Safe, healthy, sustainable food is essential to our survival, as is sane, life-enhancing work that depends upon what Lincoln named the better angels of our nature. The practices we describe will ameliorate the problems of global warming, but by themselves are not the cure. Humans need to take honest stock of their energy uses, and not wait for governments to make the fearless choices that will be required. Those decisions rest upon each of us. The time to turn down the thermostat is today.

The individual has to see the delusions of technology for what they are, and accept that our future holds no free lunches fed by fossil fuels. As biologist and environmentalist Barry Commoner pointed out thirty years ago, to his way of thinking the ideal engine for our planet, with a truly efficient and appropriate energy transfer, is a sheep eating grass to make wool. The heat exchange is minimal, and except for methane, the outputs are all beneficial. By contrast he noted how, quite apart from other environmental impacts, nuclear reactors and coal-fired power plants harbor innate and glaring inefficiencies.

In other examples that touch agribusiness-style thinking, Commoner pointed out how by the 1970's laundry soaps had been systematically replaced by detergents, due mostly to the fact that detergents generated a larger margin of profit. Despite the fact that soaps work fine, and are made of environmentally friendly ingredients—lye and fats— rather than detergents' synthetic and non-renewable resources. The shift from natural cotton and wool to synthetic fabrics is a nearly identical story. In each case, with the aid of an advertising blitz, the high-tech energy-intensive solution has all but displaced the low-tech natural one.

There are no quick fixes for our energy dilemma, no clean alternatives that will let us simply "plug and play" with the same greedy toys. And we will not be helped by the natural, understandable instinct of both business and politics to tell the listener, potential consumer or voter what he or she wants to hear. As consumers we are regularly lied to by experts, by people wanting to sell products and processes without revealing their true costs. Whether fed by ignorance or willful deceit, the effects of such deceptions are the same—to continue propelling us along the same disastrous course toward a precipice that offers the choice of radical change or catastrophe.

One eternal farming seduction involves the distinction between labor-intensive and labor-saving processes. Where we suspect we are being offered a false choice, it might be better at the outset to look into the nature of the work. If the job is meaningful and carries intrinsic rewards, then there may be little need to rescue the worker from it. And if it strikes the worker as meaningless, repetitive drudgery, no external incentives may be enough to counteract the detriment to the worker's self-worth.

We know where to start. The truth is there is no food safety or sanity without partnering with farmers, visiting the sources in the field. At the same time, the need to speak and act in the moment should not drive us to be righteous or shrill. We can imitate the innate modesty and good humor of traditional farming in this regard. There is no livelihood more grounded in reality from one season to the next, from one crop to the next in the dance of rotation, one flock or herd to the young of the season to come.

Farm Romance

This is easier work with a partner, that can spin in an instant to play. Handholding, lifted and lifting. As you share your generation's grasp on its living, along the way comes that search for the right person, one who understands your craving for good soil and strong animals, who is likewise touched by fecundity sprung from a well-tended field, what percolates to the surface and leafs out, what slows and thickens and deepens the reach of the love. Someone who like you wants to root something well, see it grow. Who knows too how the earth has to rest, sometimes more than a winter, be left fallow sometimes a whole season unfurrowed, seeming as if it had lost its way. Watched over by the two of you, who under the drifts conjure spring up ahead.

Toehold

We need to get more folks back on the land, living and working there. While we're passing around vast loans to larger players, we might consider offering new homesteads, on suitable portions of abandoned military bases and other urban and rural surplus government lands, that might be tilled without doing violence to what's already growing there. During times of economic downturn we could even offer urban and suburban lots in foreclosure as homesteads. Unlike the homesteading of the nineteenth century, which mandated clearing and often questionable "development," we might require soil conservation and replenishment as incentives to the new small farmer growing crops and raising livestock. Young people need more than the urban paycheck with its isolation and anonymity. They need purposeful effort that engages them in community, and more, they need a toehold, a stake in the future.

Handed On

There is always a delicate dance in holding farmland and handing it on intact to young farmers of energy and vision. As the longer arc of the race is reaffirmed, each generation has to find its own place, pause to dream afresh and start anew. Whether this happens on the old home place or not, it needs to happen. And the handoff is not always smooth; the previous generation may not be ready to let go, or may be forced by circumstance to release its grip before the next generation is ready to take up the venture. Seldom are there too many heirs who all want to farm, who would subdivide the home place out of existence. The real trouble comes where there is a break between generations, when one of the next generation dismisses all but the cash value of what has gone before, craves to take the quick profit and move on. Log off the woods, sell the breeding stock, find a developer for the frontage fields, liquidate the assets.

It can be hard to argue with any need that is strictly reducible to dollars and cents. But how is such a payout to be arrayed alongside the farming life it may dismantle, and move forever out of reach? We need

One Seed

to be careful to keep the wherewithal intact, the fertile working soil and its tool kit, even a bit care-worn and rusty, by offering it under modest terms to folks who will make the proper use of it, shoulder the essential good of feeding us. On our legislative wish list we might even include property tax relief for family farmers, in return for carrying on a sustainable and productive link to the land.

Sometimes it is a matter of negotiation among heirs, or a matter of striking a working deal with a widow or widower, making possible a life tenancy. Such a retirement in place might create opportunities for teaching and mentoring the next generation. There are no easy answers to this age-old dilemma. Many a man has refused with good reason to follow in his father's footsteps. Yet as one of Edwin's sons likes to say, the land wants to stay in the name.

Foresight

Speaking of sharpened tools and widening circles, in a shed by the road that runs through Edwin's farm there is a Fairbanks-Morse scales built into the floor. The scales has been weighing their produce, and arbitrating neighbors' transactions of feed and livestock for nearly a hundred years. It was bought as a kit in 1913 by Edwin's father James, and cost $190 then, plus half a day for the team and wagon to pick up from the nearest railhead a dozen miles off, then another day or two of work to assemble. Though the platform planks have been replaced, that balance-beam scale with its moving weights is still accurate and in use today. Since 1913 the family has been able to know the precise weight of every cow or hog they took to market, every bushel or bale they bought or sold. Their neighbors have also long made regular use of the scales to measure their own transactions. The scales continues to be a tool of empowerment, a check on the one-sided market of take-it-or-leave-it, and a constant nod toward community.

While we're admiring those scales, one of Edwin's sons points to charred planks in the wall, and recalls how this shed was built to shelter horses and mules after the barn burned in 1900, the year before Edwin was born. Once attention is turned to it, I catch a faint whiff of that fire over a century old. The shed was thrown up quickly, of whatever was spared by the fire, and whatever was lying around. Maybe a temporary fix until the new barn could be raised the next year, but here it still is: that moment and its aftermath embedded in the life of the place, in the thinking that even the slightest things have to be built to last, yet in a moment let go.

In the Open

A good deal of the modesty of farming may come from the sense of a life lived in the open, not constrained or channeled by other human efforts. In the field no rows of houses or stacks of rooms or ribbons of concrete surround us. We draw faint lines with our fencing, our dirt pathways and roads. And the scribble of work on the land is suficial, extends down but a little ways into the soil dark with decay and nutrients. We are like sailors riding self-made waves. The momentary meaning of that calligraphy is easily erased by a downpour, hailstorm or deluge, by infestation or plague, or by the random greenery that springs up over a season lying fallow. And the lessons abound. We leave a tool at the edge of a field, near the work, and next thing we know it's frozen in place. As would be any one of us so neglected.

When Emerson came to visit Thoreau at Walden Pond, some evenings in the cabin they would lean back against the walls in opposite corners, as if to make room for their thoughts. And Thoreau spoke of going outdoors to have a more serious talk, where a man's thoughts wouldn't seem so large, magnified by his manmade surroundings.

Though sometimes drowned out by the rattle of machinery or a tractor's roar, farming lets one work his thoughts, own what he thinks beyond the chuff and clatter of the everyday. With the harvest that chance comes in spades, when the time spent working with others fills with old stories swapped back and forth, with banter great and small. Moments when you learn who's hiding there alongside, even as you chance to show yourself. With work on the land as backbeat, in the pulse of its rhythms, how its moments condense and relax, your understanding accumulates, as you come to savor what matters.

One Seed

So for some is working in the open a way to deflate pretension, in effect a spiritual practice? Because in the field there can also be the illusion that all this I have grown. At the other end of the balance there is always ambition to be weighed, even when it stops well short of arrogance. Some farmers follow the plow and the herd because nowhere else do they have a deep enough sense of the difference they can make, where that difference begins and ends, where it makes a stand. What living actually means, what can be directed and sustained, given a turn to sustain us.

Then too, if art is an ordering and abstracting of what is, a reshaping of reality to elicit thought and feeling, then farming can also be an art. Animal breeding might be a kind of living sculpture, a graze and a frolic through time that seeks to approach perfection in each permutation, each embodiment. A living refinement of one's need. Likewise raising domesticated species of plants and trees that as they grow more attuned to the home place, more productive and beautiful, become more artful in their living understatement. More true to form.

The palette is nature herself, her chosen colors ratified by the seasons, the brushes contraptions of steel and wood, leather and rubber, and the scale, even miniaturized in gardening, is beyond reach of any canvas. To build and create and repair, to breed and water and feed and nurture, to prepare on an intimate scale and usher forth the life to come—what else could it be but an art.

The Arts of Farming

There is always a moment of doubt, a test of faith after planting, in that interval before the first plants break the surface of the smooth bare ground. Have I planted too deep, too shallow, have I been in too much of a hurry, left seed ground rough and dry, have the crows and ground squirrels found easy pickings? Is my seed lifeless after all? Then comes the faint poking up, leafing out, maybe spotty where the gophers have been at it, but still a blessed relief, the new green marching on.

In the hierarchy of human needs, perhaps the highest we are conscious of is the aesthetic—the artful self-expression of that irrepressible life force flowing through us. The only need hovering or perched above one's art is the spirit already rising half out of the body. In this regard the practices of farming paint one's deeds with the tangible and substantial palette of the possible. The arts of farming are subtle, intuitive and alive, cumulative accretions over time. And the ground tilled as we advance through a life endlessly reminds us how well or ill we have done.

Small farming is a livelihood dependent on factors beyond man's control. Its lessons are not just the extremes of abundance and disaster, and include endless subtle instances of modesty and wantonness, generosity and prodigality, punishment and forgiveness. Some of us are temperamentally unsuited to finding out if we have had a good or bad year in a couple days' reckoning each fall. But all of us can understand the gift of such abundance. And some of us can hardly wait to learn what the new day in the open will bring.

Paint the Life

In the search for challenging and bracing images of farming, we might visit Van Gogh's painting "The Potato Eaters," which looks beyond illusions of choice, to touch the core of satisfaction in the rural life. With five country people around one dish on the table, a heap of whatever is in season, the only choice is to eat or not to eat—effectively no choice at all. Yet the faces in the lamp-lit room are not sullen, angry, disappointed. Two men, two women and a girl sit dipping their forks into the common dish, their glances a web of connections that surround the simple meal, the tea being poured alongside. It is a silent moment, yet a warm one. Each person gestures with the same knotted, hardworking hands. Without any jostling or contending, the family alert to one another's comfort, the one dish seems enough. As Vincent himself said of the piece: "I wanted to convey the idea that the people…used the same hands with which they take food from the plate to work the land, that they have toiled with their hands—that they have earned their food by honest means."

One Seed

Poking Through

Ash Wednesday's come and gone. It's still chilly, with a dusting of snow overnight, though given the sunshine this morning in the still air the cattle have taken to the field. Humped over in their matted winter coats, nosing and pawing for any little bit of the new green poking through. We still have hay, though it's beginning to run low. This is a good sign they're not going to stand around in a huddle waiting to be fed. A sign too that winter won't last.

Size Matters

The issue of scale is a tricky one, attempting to find a middle ground in the mind's eye, one that shuns vast reaches and remote-controlled grandiose methods, yet avoids oversimplification in the name of virtue. We know some notions on too small a scale leave us vulnerable. We should not expect any farmer to work his land with spade, rake and hoe, as did young George Washington Carver, or that every farmer have his own roadside market, regardless of location or traffic flow. Yet every small farmer needs to seek new ways of marketing, set up his stand alongside others, and find ways to value and share what he grows.

Finally, size matters first and last, because young farmers starting out can't buy a seat in the game. Never mind the active discouragement of a quick-buck culture, without access to a large tract of land and substantial capital, or failing that, deep pockets of credit, there is no getting into farming at the scale it is currently envisioned by the powers that be. Small wonder that the average age of agribusinessmen continues to climb past retirement. Their game has no future.

By now you may have wondered if we were ever going to define who and what a small farmer is. More likely you might have thought we were going to avoid finger-pointing and name-calling, and duck the issue entirely. But if you made it this far you know that a small farmer is not to be measured by some arbitrary number of acres planted or grazed. And not by the tonnage or horsepower of his equipment, though this would begin to suggest the scale of his effort. A man who has five acres with three buildings on it housing 80,000 chickens is probably not a small farmer. Chances are that he's a specialist with a lucrative long-term contract in his pocket, an agribusinessman. By contrast, the small farmer may be identified by the modesty and diversity—and deftness—of his approach, by how committed he is to finding some mythical better way, that avoids doing violence to his own or the land's qualities. You might tell this man or woman in passing by a certain directness about some subjects and elusiveness about others. A dreaminess in the eye certain times of the day, that is busy taking the long view.

If you accept that farming is not a stopover on the way to agribusiness, not a calling we should desire or expect to grow out of, not a comforting dream or delusion we might wake from, in order to do something else, and less...

Then what? It could be that the boldest step you can take would be to decide that you are that small farmer, and begin to make moves that will deepen your connection and resolve.

6. The Trouble With Farming

The Final Feast

We aren't used to consciously considering what is at stake. And it would be tough to carry some of these thoughts in the forefront of our minds every day. But for the moment let us speak bluntly. If sustainability is a myth, a delusion, then as a species we are doomed. If farming cannot be done organically—that is, if the plant and animal life that depends on the soil cannot support us over time without the petrochemical inputs which are finite, costly, and of dwindling effect—then we might as well break out the seed corn for the final feast.

So let's not lose sight of the future. To prepare for farming beyond oil, we need to imagine and begin to live with alternatives. Dry-land for much of the country, with low-till practices, with mulch and ground cover to capture and hold runoff. With manure and compost fertilizers, with crop rotations and hardy seed stock, with breeds of draft animals suited to the work. Or that elusive alternative, zero-emissions tillage equipment.

We should not regard organic methods and produce as something quaint, a slightly healthier option that deserves its designer price. Sustainable means renewable, which is an absolute. In the case of farming, replenishing the nutrients that are spent in turn growing each crop. Relying on rot. So sustainable means able to be borne intact into the distant future. Means the life, one bright seed to another, carried on.

Leap into Being

In the long run everything shifts and moves, runs down. Tombstones slump and topple, meaning sloughs off in the weather. Even this full moon I have watched building over a week, that throws a cool and gratuitous light on the dead and the living, that bulges tides round the world, will not always shine, tangible as it is useless in the long run. Though for all that, seeds respond to its tug and its touch, reach up and leap into being.

Spare Change

We know there is no free lunch, no carpet to sweep questions under. Though we can playfully imagine standing in some solar-powered electric world, eyeing the drizzle and saying "It'll be one sunny day, when I get to town again," in the here-and-now we are more bluntly skeptical. When we hear that a new all-electric car will travel 230 miles on $5 worth of electricity, we immediately wonder where that electricity will come from, how and where it will be generated, what the real environmental costs might be. We know our dilemma won't be solved by a single high-tech marvel, but will come from a whole new grid erected upon countless small choices—on nothing less than a shift in consciousness.

Which makes us wonder if it's ever happened before. One modest example might suggest the possibilities. In the mid-60s the U.S. Mint faced an impending crisis. The country was running out of change. Consumer demand was outstripping the mint's ability to strike new coins. But then something funny happened. Within a few weeks and months after this story appeared in the national media, a spontaneous shift occurred. Ran through the country like a shiver. Instead of assuming that a purchase entitled the purchaser to correct change, shoppers began seeing the offer of correct change—or as near as the shopper could come—as a matter of social etiquette. And in every kind of business, little bowls of change appeared next to cash registers, with notes saying "take a penny, leave a penny."

Which was the answer. Free range pennies. No laws had to be written, no sermons preached. The U.S. Mint did make "sandwich" coins to ease demand on some precious metals, but disaster was averted. The problem fixed itself, and has stayed fixed.

Baby With the Bath-Water

Remember that old country injunction, "Don't throw the baby out with the bath-water," and hear how it resonates with our current farming situation. Implicit are two laugh lines worth a listen. One is that it's an uncommon baby so placid in the bath that it could let itself be discarded. The other implies a level of uncommon carelessness, how bath water could ever get so dirty that it could hide any baby, and not have been changed long before.

So it's high time we paid attention to some small good-natured players that have been all but overlooked. And to some old dirt, that has obscured what needed doing. And should you find yourself in the bath, you need to make a fuss.

This little book has been an attempt to lather up and rinse the head, then comb some of the snarls and tangles out of our thinking. It holds inklings of some new and old ways of farming run in parallel, that seek to add up to encouragement. It has been an attempt to speak to the urgencies of the moment as well as to the timeless verities. Both ways there is too much at stake. Implicit too is a whiff of hope. We need to study the wisdom and ways of the past for what they still have to offer, all the while sifting through the latest solutions for the sturdiest and most subtle and long-lasting. I'd like to think it is not so much offering ammunition for combat, as help in making choices, alternatives to chew over and digest.

To conserve is to protect from harm or loss. Small farming is conservative in the best, most ancient sense, rescuing our resources of soil, seed and species by means of its practice and values. By the same token, small farming is radical, rooted. And fundamental, based on the untouched landscape. Small farmers achieve such enduring effects, they can't afford to be dogmatic, hidebound, humorless. There is after all affection and play in the work. And a better way of doing things might lurk just over the horizon, that deserves testing alongside the best current practice.

This is a small book, to conserve time in the reading, resource in the making, space on the shelf, in the pocket. It is a place-marker, a brief about scale, a start, not a finish. As befits its subject, a work in progress. The windows here that look out on the farming life could as well be your own.

Many of the questions and meditations touching farming reach beyond it, employ that life, its arts and strategies as a metaphor for the life fully lived in any guise. The debilitating effects of agribusiness on the rural scene go hand in hand with increasing corporate domination of the whole culture. We have all been branded as consumers — belittled, denied meaningful work, swept aside. Yet farming remains a primary source of our human connectedness. It provides a fine metaphor, articulate in its details, worldwide in its implications, intimate and humble in how it touches each of us. But as a need and a practice it is also real. And we are in it together.

What I Feed Myself

Early last summer I spotted a farmer parked in the gravel at a crossroads, and pulled over. He leaned by the tailgate in sunburn, straw hat and overalls. Long sleeved work shirt given two turns at each wrist. The back of his pickup was full of baskets and crates of fruit. A sign read "Yakima Peaches." I got out and walked over to study those peaches, which seemed like they might be a bit early. I hadn't yet touched one when he opened a jackknife, lifted a peach from the pile I was looking at, cut a wedge and held it out to me on the point of his knife. All without a word. And while I ate it, he polished off the rest of the peach, leaning over to keep the juice from getting all over him. Then pulled out a handkerchief to mop his face. No question who grew those peaches. It was a sales job that would not have looked out of place hung up alongside the Last Supper.

In Plain Sight

The trouble with farming is it's simple and direct. And anything but. An insistent waiting game, daily and hourly, leafing out minute by minute. It's the demeaned and discarded identity, the calling nearly everyone will answer once the belly starts to grumble. It's ancient, timeless, what else but the latest way we've thought of trying to keep fed. It's unforgettable but dismissible, even growing up right in our faces. It tries to lift and not drop us, tries not to get us poisoned, tries simple carrion comfort. It's practically always in plain sight, unfurling its thoughtless arabesques. It will only get lost if we neglect it. In any number of ways in all seasons it will feed us. Even sharing covers in the cold night, it troubles and savors our rest.

Paul Hunter has worked corn, hay, wheat and oats, from plowing to harvest. Stretched fence, raised cattle and hogs, forked and spread manure. Tended the full range of whatall goes in a garden. Fixed machinery. After a good while spent teaching high school English, history, arts and manners, he mostly writes, whittles, sets type, and plays music. He has three interwoven books of farming poems from his Indiana upbringing, with more on the way. A Seattle teacher, performer and lecturer, he serves on the board of the newly founded Small Farms Conservancy, and invites you to join in its work.